An Introduction to Molecular Modelling, from theory to application

Dr Dimitrios P. Vlachakis

B.Sc., M.Sc., M.Phil., Ph.D.

ISBN Number: 978-0-6151-7606-2
University of Cardiff, Wales, UK

Table of contents:

Chapter 1: Theory

1.1 Molecular Modelling

Molecular modelling is very useful for investigating, comparing, analysing and visualizing chemical structures and for giving qualitative and quantitative information about biological systems.

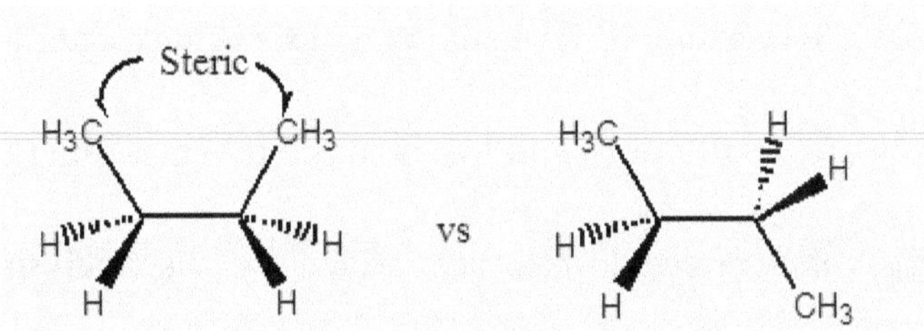

Figure 1. Steric hinderance of a small organic compound

Figure 1 shows a characteristic example of steric hinderance. Two dimensional models like this only contain qualitative information. Quantitative information can arise through molecular mechanics and in conjunction with a computer, where the physical properties of the molecules can be evaluated and analysed based on a set of predefined criteria concerning various chemical properties (such as bonding, charges, steric hinderance…). Molecular Modelling can be used to study the geometry, the energy and the chemical properties *in silico* so efficiently that nowadays it is possible to predict the outcome of chemical reactions, design reactions, determine the unknown three dimensional structures of proteins, screen and design new and effective drugs. In this part, the basic principles and theory of the methods that have been used in this chapter will be presented.

1.2 Molecular Properties

The geometry and the overall structure of a molecule are described by its bond distances, dihedral angles and bond angle. This unique set of angles and distances create a set of coordinates that define the positioning of each atom in that molecular structure in three dimensional (3D) space. The energy condition of this molecule can also be assessed and evaluated. The energy of a molecule includes all forms of energies, such as kinetic motion (described by vibration, rotation and translation) and forms of the potential energy of the molecule. The potential energy of a molecule can be defined by the analysis of the electrostatic interaction between charges, the magnetic interactions between spinning charges and finally the potential energy of the bonds of the molecule. The total energy is indicative of the reactivity and stability of that a molecule or a system. Below is a reaction coordinate diagram that indicates the energy changes during the course of a chemical reaction.

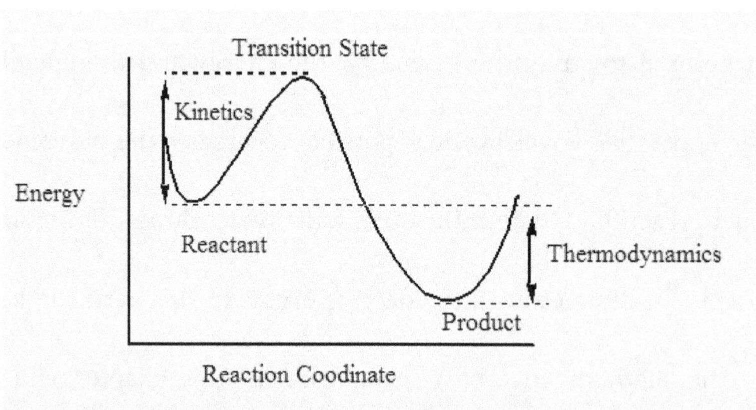

Figure 2. Energy changes during the course of a chemical reaction.

Here the products are in the lowest or global minimum, the transition state is at energy maximum and the reactants are at an energy minimum. The dotted lines in the above diagram are indicative of the reactivity of the system (its kinetics) and the thermodynamic stability of the system. Through molecular modelling it is possible to quantify the above characteristics of the system and, for example, predict its reactivity. There are two fields in molecular modelling that attempt to do this: molecular mechanics and quantum mechanics.

1.3 Molecular Mechanics & Forcefields

Molecular mechanics are based on the ball and spring representation of molecular systems. Here, the atoms are considered to be little balls, with varying properties according to the element, and the bonds are considered to be the springs that make the two interconnecting balls interact with each other. The ball and spring model is described by Hook's law, which evaluates and quantifies the energy of the stretching of the spring.

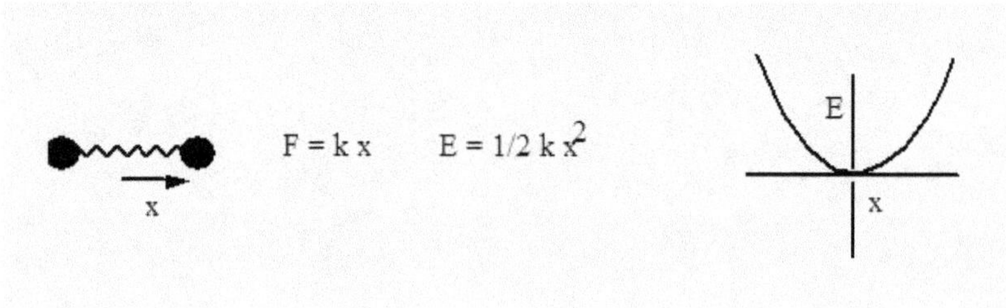

The force constant is the constant k. The energy that is contained in the spring and the restoring force of the spring are proportional to the force constant. The force constant will determine the strength of the bond that the spring represents. The vibrational frequency of the spring is described as:

$$n = \frac{1}{2\pi}\sqrt{\frac{k}{\mu}}$$

The vibrational frequency (n) has been estimated to be proportional to the square root of the force constant (k) and inversely proportional to the reduced mass of the atoms that participate in a bond.

All of the above can be combined and through potential energy functions of various structural features, such as bond lengths, bond angles and non-bonded interactions, can describe a forcefield (figure 3). There are many different ways to set a forcefield depending on the needs of the system under investigation. Usually the factors affecting the energy of a molecular system (bonds, angles, dihedrals, non-bonded, etc), are evaluated separately and they will contribute to the value of the total energy of the system. The most popular forcefields are the MM2, which is suitable for small molecules, hydrocarbons and some simple heteroatom functional groups, AMBER or CHARMM, which are parameterised to be used for peptides, nucleic acids and generic macromodels.

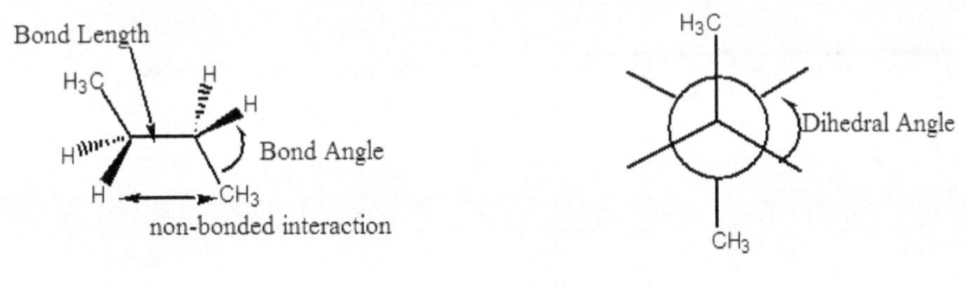

Figure 3. Total energy is affected by bond distances, bond angles, dihedral angles and finally non-bonded interactions

Overall through molecular mechanics the total energy of a molecule is described as a sum of all the contributions that may arise from loss of equilibrium in bond distances, also known as stretching contribution, bond angles, known as bending contribution, dihedral angles, the torsion contribution and finally non-bonded interaction contributions.

$$E^{total} = \underbrace{\sum_i E_i^{stretch}}_{\text{bonds}} + \underbrace{\sum_i E_i^{bend}}_{\text{bond angles}} + \underbrace{\sum_i E_i^{torsion}}_{\text{dihedral angles}} + \underbrace{\sum_i \sum_j E_{ij}^{non\text{-}bonded}}_{\text{non-bonded atoms}}$$

The energy that is stored in chemical bonds of a molecule can describe the stretch, bend, and torsion energy whereas it is the steric attraction or repulsion that represents the non-bonded energy. The latter is broken down to two different categories: the van der Waals (VDW) and electrostatic interactions.

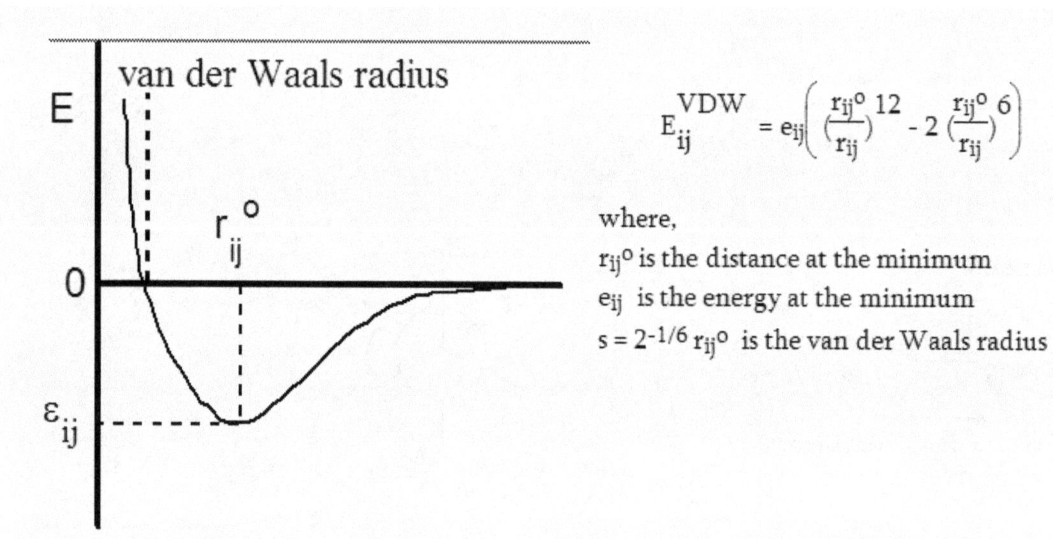

Figure 4. The van der Waals interactions plot and formula.

A very steep energy barrier is generated at the van der Waals radius of each atom. Moreover a very shallow energy well is produced at larger separations (figure 4).

The inherent steric size of atoms and elements is dictated by their VDW radii. The same metric is used to describe weak attractive forces between atoms in close proximity. A trivial example of the weak van der Waals attractive forces is the condensation of a gas into liquids. Furthermore it is the van der Waals radii of each element that is used for its visualisation purposes in space filling models of the molecule they participate. Steric repulsion takes place only in the case where two atoms come closer than the sum distance of their VDW radii.

1.4 Energy Minimisation

As soon as the set of the internal coordinates of a molecular system has been determined, computer algorithms can be used to help find those coordinates which will account for the lowest energy of the system. All bond angles, lengths, dihedral angles and the relative energy between various different conformations of a given system will be evaluated in order to determine the minimum energy conformation. It is crucial to understand that reducing the strain energy of a given molecular system does not mean that the system will reach energy minimum (also known as global minimum). An example is the following figure (figure 5) with two different conformations of butane:

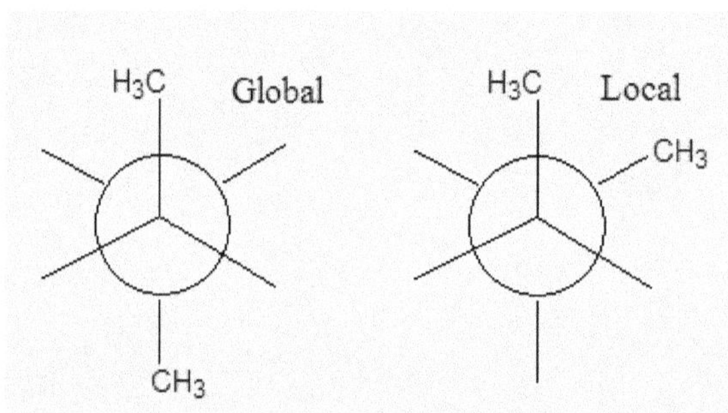

Figure 5. Two different conformations of butane

An energy minimisation algorithm will allow the rotation of groups, when their bonding allows. The rotation of the groups will give the molecule the opportunity to explore different conformations that will account for different energy values, thus allowing the compound to move towards its global minimum conformation.

1.5 Quantum Mechanics

Quantum mechanics are useful for the evaluation of electronic properties that may influence physical and chemical reactions between different groups of a biological system. The estimation of energy conformational profiles and that of intermolecular interactions in various cases can be achieved through quantum mechanical calculations. The quantum mechanics arise from the Schroendinger equation. In a very simplistic representation the latter equation can be described as:

$$H_{psi} = E_{psi}$$

Where: H is the Hamiltonian operator,
Psi is the wave function
E is the total energy of the system

Using the Schroendinger equation and the above operators a complete description of any molecular system can be achieved.

At a given molecular system the Schroendinger equation can be solved in two different ways. First is the no approximation or *ab initio* approach. Secondly, the Schroendinger equation could be solved by the introduction of some approximations, which is a method also known as semi-empirical approach. The main advantages of *ab initio* methods is that all electrons are explicitly included and they do not requiring any specific parameterization, thus making *ab initio* calculations a universal tool for all molecular systems. On the other hand in semi-empirical calculations it is the valence electrons that are explicitly included. Here most of the integrals are neglected or approximated by a set of parameters that arose from experimental work. Choosing the most suitable approach to investigate a given system does not solely depend on the size of the molecular system, but on the type of unique parameters and molecular properties of the system. For example, the conformation, electrostatic potential and, electron density of the given molecular system.

The most popular semi-empirical calculation algorithms are the AMI, MNDO, CNDO and MINDO, whereas for *ab initio* calculations the Gaussian program is very representative. AMPAC and MOPAC on the other hand are very widely used for *ab initio* calculations.

1.6 Molecular Dynamics

Molecular dynamics simulations are used to describe the patterns, strength and properties of drug-receptor interactions, the solvation of molecules, the conformational changes that a molecule may undergo under various conditions and other events that require the systematic evaluation of molecular properties in dynamic molecular systems. Molecular dynamics are concerned with the motion of atoms and molecules. An example is the normal mode analysis, where the dynamic motion of a molecular system is evaluated from its total energy. According to Hook's law, the total energy of a small and simple diatomic molecule should be given by:

$$E(r) = \frac{1}{2} k(r - r^o)^2 \quad € \quad F(r) = -\frac{dE}{dr} = -k\,(r - r^o)$$

Where: E = energy, F = force, r = distance, r^o = initial distance, k = Hook's constant

Here, the forces on the atoms are estimated by the derivative of the energy. When the forces have been assigned the Newton law of motion can be used to solve the molecular motion (F = m x a, force = mass x acceleration). In the case of the small and simple diatomic molecule the displacement from the equilibrium bond length (x) will be given by the formula:

$$x = r - r^o \text{ and } \mu = \frac{m_C\, m_O}{m_C + m_O}$$

Where: μ is the effective mass of the vibrating diatomic molecule.

In molecular dynamics simulations the kinetic energy of the system will depend on the temperature of the system. The total energy will be the sum of the kinetic and potential components of the system. The acceleration of each atom is estimated from the set of forces it accepts, under the given forcefield. The results generated can be used to estimate the configurational and momentum information for each atom of the system (i.e. energy, pressure). Molecular mechanics therefore can be used to further optimise a model generated by homology modelling as well as docking results, where the protein / ligand interactions can be analysed for their stability during a specific time.

1.7 Homology Modelling

Homology modelling is used in order to predict the 3 dimensional structures of proteins with unknown 3D structure, using solved homologous proteins as templates.

Homology modelling claims that the biological structure of a protein is more related to its biological properties and function than its sequence. A homologous protein is a protein that belongs to the same family, has the same function and shares more than thirty percent similarity with the protein of interest.

The first step of a homology modelling algorithm is to set up and optimise the sequence alignment between the query protein and its template. Sequence alignment is broken down into four steps. Firstly, it uses rapid alignment methods to calculate all pairwise similarity scores. The second step is the generation of a similarity matrix. Then the sequences are clustered according to the generated similarity matrix with the aid of an algorithm. The next step is the generation of a cluster alignment using a consensus method and finally a multiple-progressive alignment is generated. The groups of the sequences are aligned according to their cluster branch order.

After that the algorithm will perform an initial partial geometry alignment for the sequence of the template protein with the unknown structure. The initial geometry will be copied from various regions of one or more template proteins. If there is residue identity, between the alignments, then all coordinates are copied. That includes backbone and sidechain. If there is not residue identity but still residue similarity is retained, only the coordinates from the backbone atoms will be passed on. In cases of zero identity or similarity a gap will be left on the model, which is also known as loop. A loop will be modelled by borrowing coordinates from any protein (from the Protein Data Bank) that matches the required sequence. The sidechain is generated automatically using a build-in rotamer explorer module.

Finally, the new models must adequately meet and satisfy a scoring function that ensures that the degrees of the non-polar sidechain groups that are buried are within range and that all hydrogen bonding capabilities have been explored.

1.8 Model Evaluation

The N-Cα and the Cα-C bonds in a protein are capable of rotating. The rotation of these bonds can be described by the phi and psi torsion angles (figure 6 – bottom right). The values of these angles define the secondary element that each set of residues will form. To visualise phi and psi angles in a protein a Ramachandran plot is constructed. Psi angles are found in the YY' axis, whereas phi angles are found in the XX' axis (figure 6).

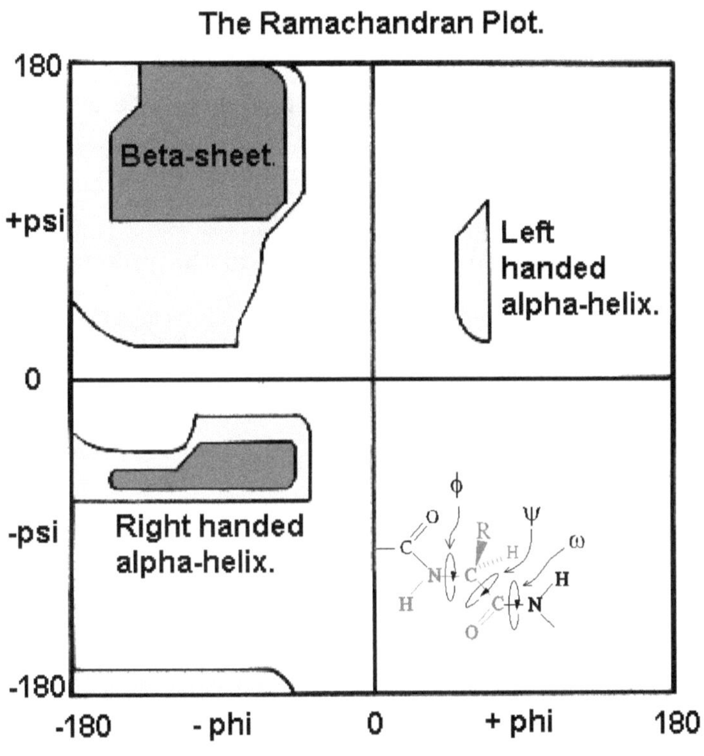

Figure 6 The layout of a generic Ramachandran plot. Indicating the major secondary element areas of a protein. Bottom-right: a typical residue with the phi, psi and omega angles designated.

The white space on the diagram above describes the pairs or groups of atoms that have distance smaller than the sum of their van der Waals radii, because of the conformation of the backbone. As a result a sterically unreal conformation will be generated, thus making the combination of those atoms disallowed. The area with no steric clashes is represented in red color. The residues in the red area are considered to be in the allowed regions and this is where the alpha helices and beta sheet conformations are usually found. The yellow areas include the pairs or groups of atoms with radii a fraction less than the sum of their van der Waals radii. This is the area that the motif of the left handed alpha helix would be found, where the atoms are allowed to come a little closer. The generously allowed areas of the Ramachandran plot are areas with pairs or groups of atoms with radii quite less than the sum of their van der Waals radii, but not significantly clashing with each other. Glycine is a versatile amino acid, because it does not have a side chain. Glycine can take phi/psi angles in each of the four quadrants in the Ramachandran plot. As a result glycines are usually encountered in loop regions in the protein, where it would be impossible for any other residue to be, because of the steric hindrance.

1.9 Molecular Docking

The last molecular modelling technique that is extensively used in this project is molecular docking. The docking algorithm is basically split into two main parts: the searching algorithm and the scoring algorithm.

The searching algorithm will explore all conformations of the ligand within the space available. Practically, it is impossible to perform all these calculations for every compound so most of the rotational and translational states of each compound will be explored within a given threshold of identical conformations. Each compound is not a rigid body, but is a dynamic structure that exists in an ensemble of different conformations. The user can define how fine the docking algorithm will be by altering the various parameters of the task. Very fine calculations are much more accurate, but also much more time consuming. The most popular docking algorithm approaches can involve a coarse grained molecular dynamics simulation or a linear combination of many structures or a genetic algorithm that generates new conformations as it moves along.

The second feature of the docking algorithm is its scoring function. The scoring function must be able to accurately evaluate each different conformation using certain forcefields and rules from physics, and return a value that will describe the energy of the system at the given conformation. Low energies indicate better, more stable interactions.

1.10 Drug design

In the last few years the number of biological targets suitable for drug discovery has increased enormously and, at the same time, computer-aided drug design techniques are becoming more popular everyday, sustained by the need to accelerate the drug discovery process. Structure-based methodologies are now widely used to design and optimise new potential drugs and, in particular, *de novo* approaches (structure based novel inhibitor design) remain very attractive to researchers as a time and cost efficient methods to design novel entities, despite the challenges associated with them.

Basically, three questions have to be addressed by a *de novo* design program: how to assemble the candidate compounds; how to evaluate their potential quality; and how to sample the search space effectively.

In answering the first of these questions, we can consider two main approaches to build the desired structures. The first (figure 7-top) is the linking procedure, where the algorithm selects the most suitable moieties to interact with the active site of the protein and then starts to link them together in a chemically appropriate way. The second approach is the growing procedure (figure 7), which involves the determination of a group as a starting point and then the growing of a larger compound that would fit in the active site and that would be capable of establishing interactions with it. The first approach requires that the user must supply the docking/interaction points of the moieties that will be used as anchors to the final structure.

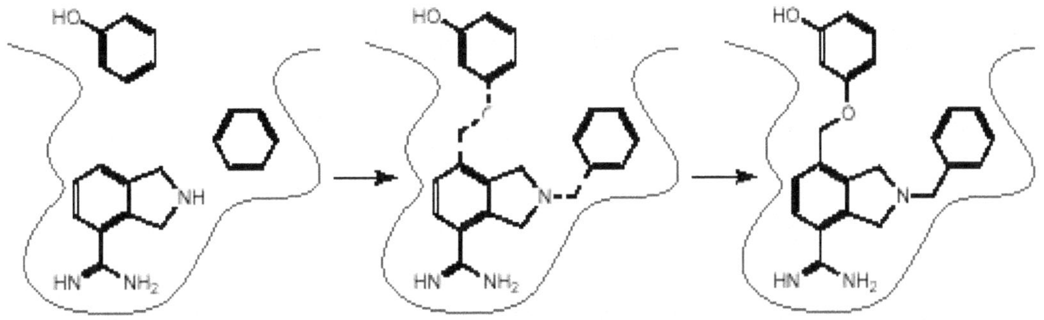

The linking approach

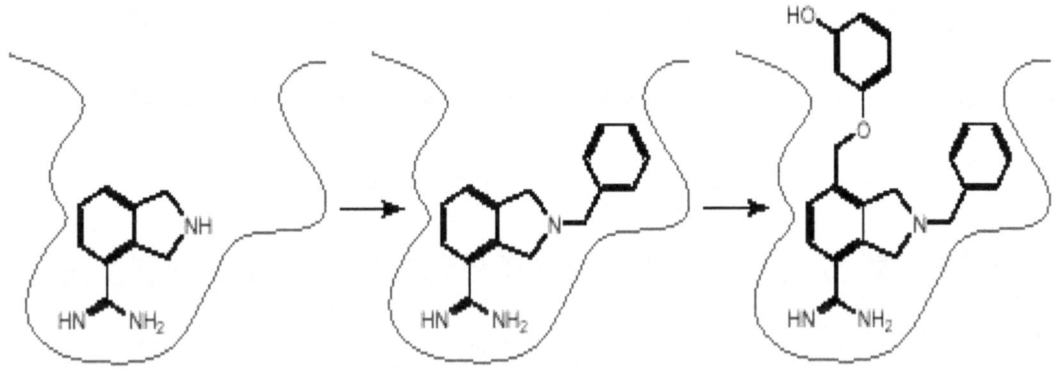

The growing appoach

Figure 7. LigBuilder's approaches to structure-based drug design.

The second approach requires that the user selects a suitable "seed" as a starting point, and that the user directs the growing process. The best results will be obtained from the combination of these two procedures.

Evaluation of the results obtained could be done by different methods: receptor-based scoring functions: explicit force-field methods; empirical scoring functions; and knowledge-based scoring functions. All of these approaches attempt to approximate the binding free energy. Force fields are computationally more costly than the other two types of scoring functions.

The last issue related to the *de novo* drug design methods is the sampling of the chemical space associated with the growing compounds. There are many different search algorithms implemented in the variety of software programs available, and new ones are continuously developed.

It is also important to mention the nature of the "building blocks" used to construct the structures: some programs connect single atoms, others use a library of chemical fragments. The size and the diversity of the fragment library is fundamental for a broad search of the chemical space. The main problem associated with the *de novo* methodologies is represented by the chemical feasibility of the structures generated: unfortunately, computers do not know organic chemistry, yet.

In our project, we have used a *de novo* approach for the design of some helicase inhibitors using a software called LigBuider. This package can use both the "link" and the "grow" approaches and implements a genetic algoritmm based search for the determination of the best structures. Once the compounds are drawn from the population, the new population is formed by combining the information from groups of selected individuals. This is done by randomly splitting the representative strings of the parents and recombining them to form new entities. This operation is called Crossover. Because the parents have been selected for their fitness, the children of these parents will, statistically, be more fit than the parents. This is done until a new population has been built from the old one. The final step is the evaluation of the new population, which may result in starting the whole process over again by changing a single or groups of parameters. The cycle described in figure 8 is continued until a solution that is good enough for the experiment's criteria has been obtained.

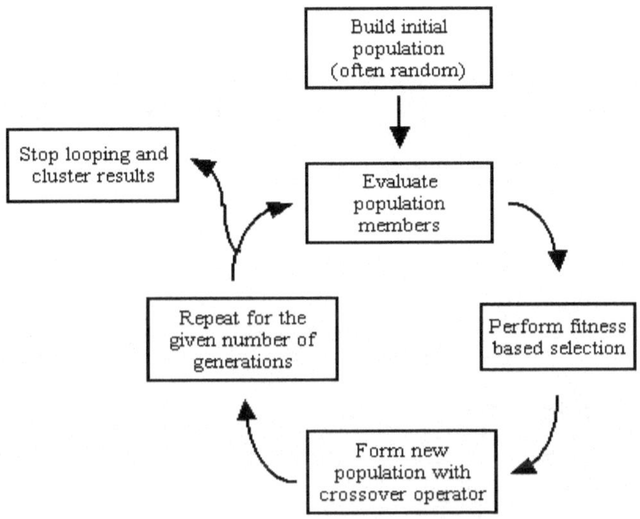

Figure 8 The steps of the genetic algorithm of LigBuilder

Chapter 2: Application
(Example)

Molecular Modelling of the β-ketoacyl carrier protein synthases.

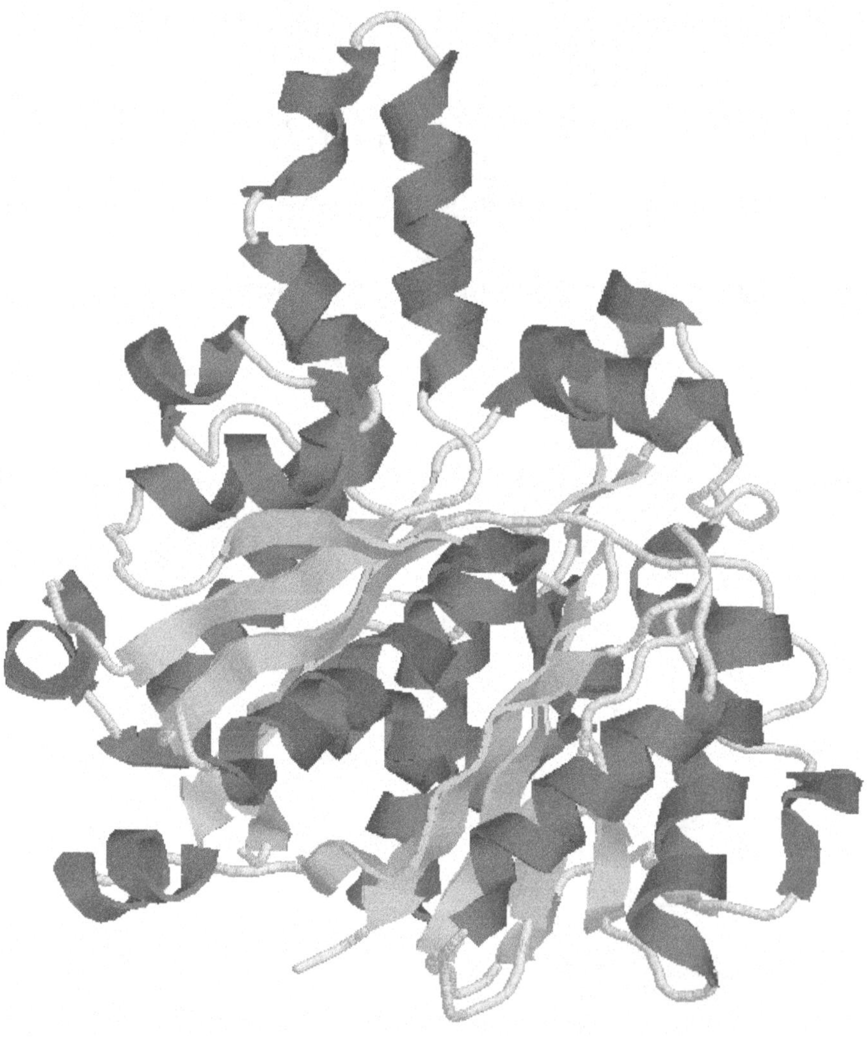

2.1 Abstract:

This project consists of a study of the β-ketoacyl carrier protein synthases (the KAS enzymes). The ultimate aim is to study the Plasmodium enzymes as potential drug targets and to establish the structures of the KAS I/II and KAS III enzymes from their sequences. In bacteria, three KAS enzymes have been identified (KAS I, KAS II and KAS III), whilst in Plasmodium a KAS I/II and KAS III enzymes have been identified. The first step was to study the active sites of enzymes from different sources (i.e. derived from different species). The main aim of the active site study was to identify the residues that comprise the active site, the ones that actually interact with the ligand and finally to identify what kind of interaction takes place when the inhibitor is docked.

The protein has a total of four active sites, which were found to be different to each other rather than four copies of the same active site. The active sites differ not only in the type of interaction that they establish with the ligand, but in the case of Cerulenin as a ligand, the active sites of the KAS I/II enzyme differ in the number of the residues involved in the ligand protein interaction. That is very interesting biochemically, because those differences between the four active sites imply that the affinity of each active site for binding to the ligand might be different as well.

The KAS I/II and KAS III sequences were structurally aligned against a variety of different sequences of the same proteins from different species. The sequences of the proteins that were used to align our unknown sequences were obtained from large web databases, and the alignment was performed using a variety of different sequence aligning packages, to make the alignment more reliable. The alignment was checked by hand too, to make sure that specific residues that were evolutionary conserved and usually part of the active site of the protein were aligned correctly. Then by homology modelling it was possible to construct a model of the KAS I/II and KAS III Plasmodium enzymes. Again homology modelling was performed with more than one software package, to make sure that the optimum model was obtained. Finally, the quality of the model was assessed, and its accuracy and reliability ware confirmed.

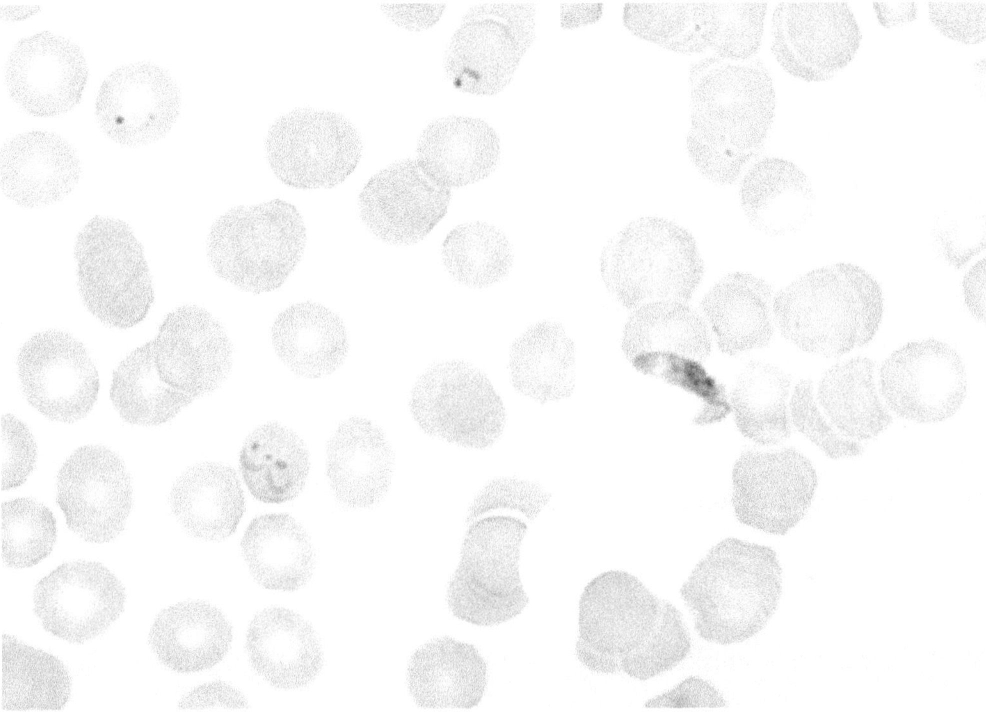

Figure 9. *Plasmodium falciparum* malaria, potentially fatal and often drug-resistant, in a peripheral blood smear. Multiple ring forms in the left portion, which can parasitize up to 60% of the total circulating Red Blood Cells. Banana-shaped gametocyte is seen at the right. The photograph is courtesy of Dr. Richard Feddersen.

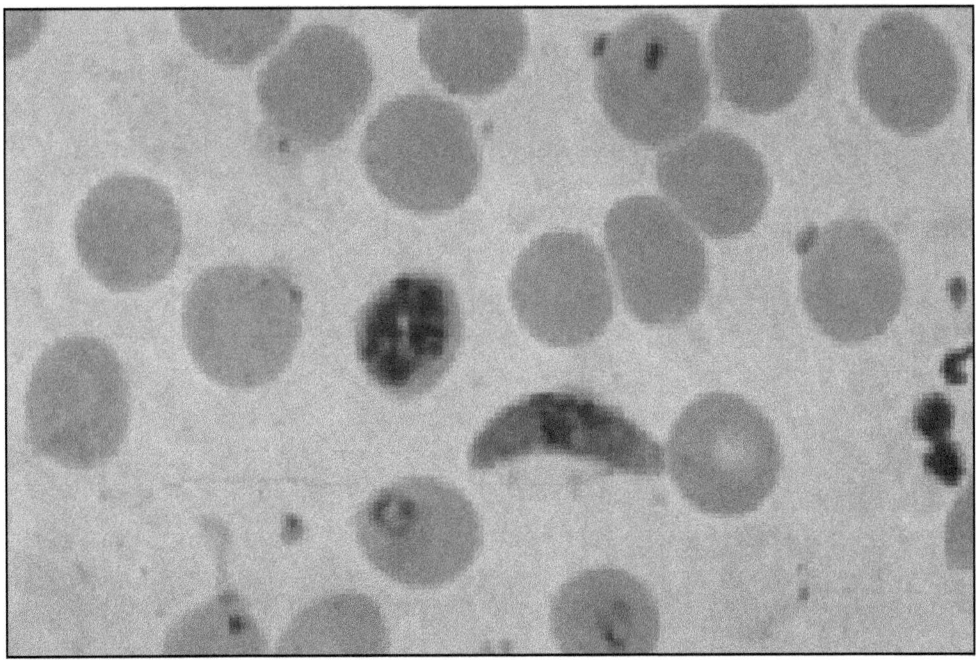

Figure 10. Microscopist's view of *Plasmodium Falciparum.* The photograph is courtesy of Dr. Richard Feddersen.

2.2 Introduction:

Malaria is a parasitic disease transmitted by mosquitoes that can be lethal if it is left untreated (figures 1 and 2). The real cause of malaria was discovered in the end of the 19th century. It was found that the real cause of malaria is a one-cell parasite called plasmodium. This parasite can be transmitted from an infected person to another from the bite of a female Anopheles mosquito. The mosquito is after blood as it is essential for it to nurture her eggs. There is an urgent need for new drugs since most known antimalarial drugs is are rapidly losing their effectiveness as the rate of drug resistance is increasing. *P. falciparum* is a member of the protistan phylum Apicomplexa (mostly parasites, the Apicomplexa are named after a characteristic apical complex of microtubules within the cell). A curious feature of the apicomplexan parasites is that these single celled organisms harbour within themselves a plastid – otherwise better known as an organelle found in plants and algae and classified into leucoplasts, chromoplasts and chloroplasts. The association, symbiotic today, is thought to have arisen via the engulfment of a cyanobacterium-like prokaryotic cell by an ancestor of the parasite (McFadden andRoos 1998).

The β-ketoacyl carrier protein synthases (β-KAS) are a key regulator of fatty acid biosynthesis, which has emerged as a target for the development of therapeutic agents. The main role of the β-KAS is to regulate the synthesis of fatty acids. Thiolactomycin and Cerulenin are two natural ligands, which are capable of binding to the active site of the β-KAS enzymes and inhibit their function. A variety of different structures of the β-KAS enzymes from mainly different origins have already been determined from X-Ray and NMR studies. An extensively studied source of proteins is the E.Coli bacterium. E. coli (short for Escherichia coli) is a type of bacteria commonly found in the intestines of animals and humans. E. coli, the whipping boy of microbiology and genetics labs around he world, is the most encountered bacterium in the clinical laboratory. As part of the normal flora of the human intestinal tract, E. coli plays a crucial role in food digestion by producing vitamin K from undigested material in the large intestine. E. coli is a very versatile bacterium. Its main advantages of being a very strong strain in terms of survival, requires only basic nutrients and replicates very fast. As a result, a lot of research has been done on E.Coli including X-ray and NMR studies of its KAS proteins.

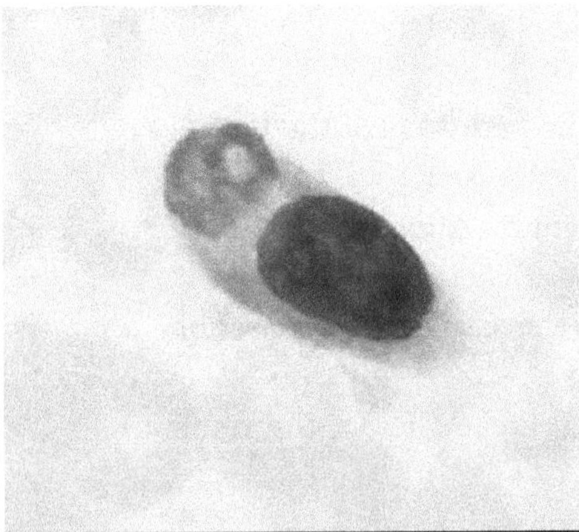

Figure 11. Malarial parasite in monocyte.

There are two main fatty acid synthesis systems. The type I system, also known as the associated system, is found in the most complex organisms, such as mammals. This system composes a unique and multifunctional polypeptide. On the other hand the type II or dissociated system, is found in lower organisms such as bacteria and plants. This system involves the cooperation and coordination of a variety of different enzymes, each one of which has a unique task in the elongation of the fatty acid chain.

Three different synthases have been found to regulate the pathways of the type II system (KAS I, KAS II and KAS III). According to the work done by Allen C. Price *et al.* (2000)[1] the KAS enzymes are mainly known for catalysing the Claisen condensation reaction. This is done by the transport of an acyl primer to malonyl-ACP. So, the result will be a β-Ketoacyl-Acyl Carrier Protein elongated by two carbon units (figure 11).

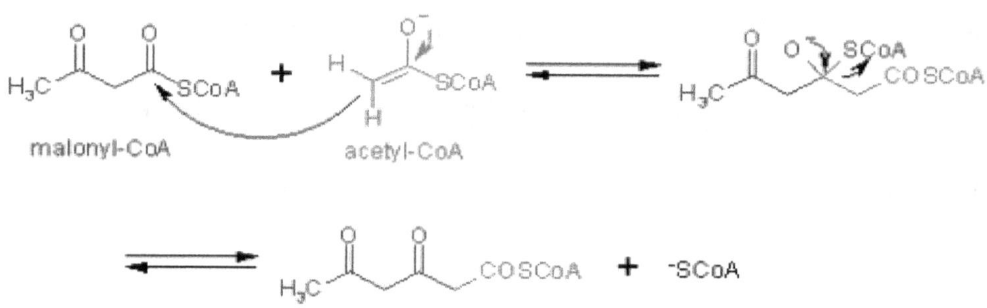

Figure 12. The Claisen reaction between acetyl-CoA and malonyl-CoA illustrates how b-keto esters are built up by Nature using the enolate derived from acetyl-CoA as nucleophile. This reaction plays an important role in the biosynthesis of fatty acids.

The β-Ketoacyl-Acyl Carrier Protein Synthase I (KAS I) plays a very important function in the elongation of fatty acids. Studies have shown that mutants lacking the KAS I enzyme face serious problems in growth and exogenous unsaturated fatty acids are required. The function of KAS II (β-Ketoacyl-Acyl Carrier Protein Synthase II) is mainly to control the temperature dependent regulation of fatty acid composition. Lack of KAS II will result in failure of the elongation of palmitoleate to cis-vaccenate. Even though under standard culture conditions growth is not suppressed. Finally KAS III (β-Ketoacyl-Acyl Carrier Protein Synthase III) is responsible for controlling the rate of fatty acid synthesis. It actually catalyses the first step in the pathway.

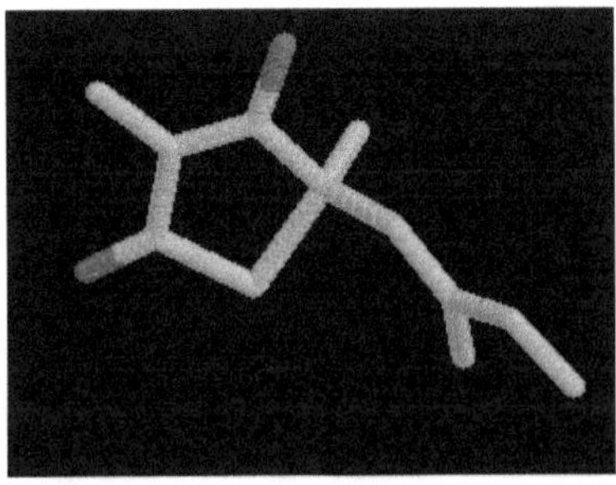
A
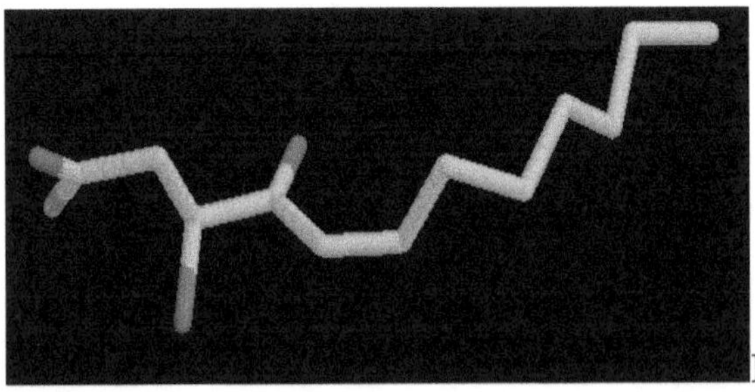
B

Figure 13. The structures of Thiolactomycin (A) and Cerulenin (B).

The β-ketoacyl synthase (KAS) components of fatty acid and polyketide synthase create new carbon–carbon bonds by carrying out Claisen condensations. There are three main steps in the reactions that result in the elongation of specific primers. The first part involves transfer of an acyl primer substrate from acyl carrier protein (ACP) or CoA to a cysteine residue of the condensing enzyme. In the second part occurs decarboxylation of an ACP-bound elongating substrate to give a carbanion; and finally the last part is the condensation of the carbanion with the carbonyl carbon of the enzyme-bound acyl primer. Three KAS isoenzymes participate in synthesis of the fatty acid carbon skeletons in plant plastids and *Escherichia coli* as part of multifunctional fatty acid synthase complexes in which each protein is encoded by a discrete gene. Substrate specificities of KAS I, II and III overlap. KAS III is specialized for the first elongation step from C_2 to C_4

and uses a CoA-activated primer substrate. KAS I and II employ ACP-activated primer substrates. In plastids KAS I extends C_4 to C_{16} whereupon KAS II carries out an additional step to give C_{18}. KAS proteins with 85% identity to KAS II, called KAS IV, in the seeds of some plants are implicated in synthesis of intermediate chain lengths for C_8 to C_{14} [1,2]. Classic studies in *E. coli* suggest that KAS I is unique in carrying out the first step of the unsaturated pathway whereas KAS II is unique in accomplishing the last step [3]. In both plant plastids and *E. coli*, KAS I and II sequences show extensive homology, whereas KAS III sequences are more closely related to those of the polyketide chalcone synthase (CHS) enzymes [4,5]. Disparities in substrate specificity are also reflected in differing sensitivities to antibiotics; for example, with cerulenin, which binds covalently to the active-site cysteine, KAS III is insensitive, KAS I is highly sensitive and KAS II is moderately sensitive.

In this experiment mainly the β-Ketoacyl-[Acyl-Carrier-Protein] Synthase I from Escherichia Coli was used. That includes the Synthase without any ligand docked, the Synthase complexed with Thiolactomycin (figure 4) and finally the Synthase complexed with the natural product drug Cerulenin (figure 5).

The structure of the β-ketoacyl carrier protein synthases is a dimeric structure with each of the monomers made up with a duplicated helix-sheet-helix motif (figure 6).

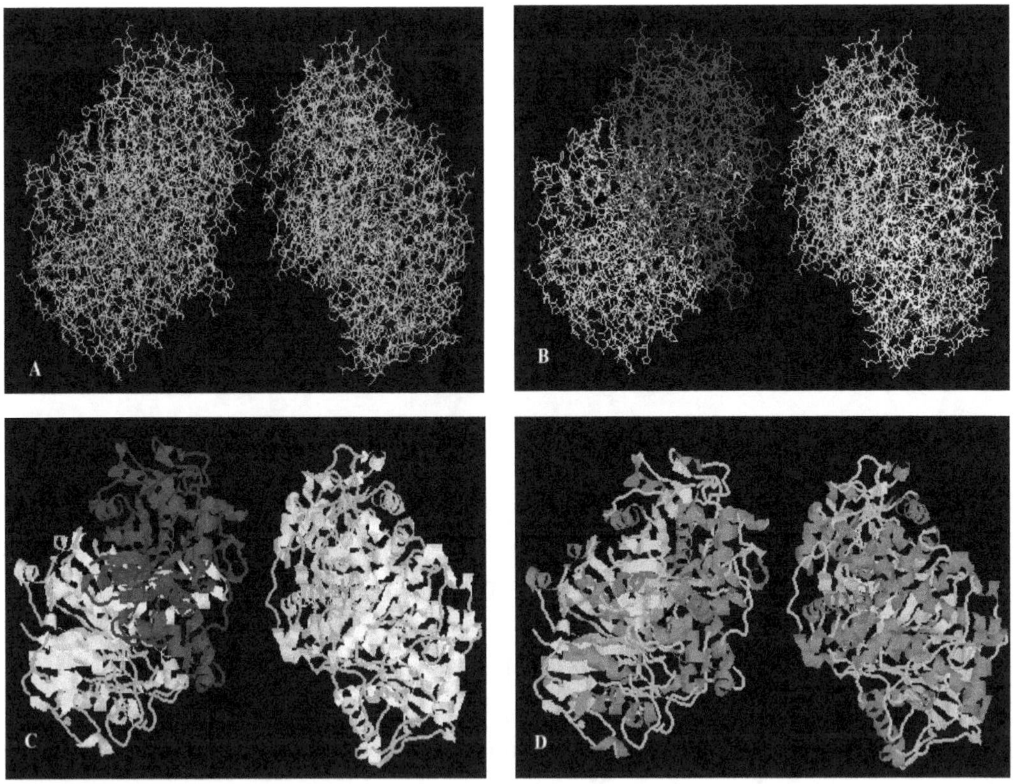

Figure 14. The structure of β-Ketoacyl-[Acyl-Carrier-Protein] Synthase I. Coloured in CPK (figure A), coloured by chain (figure B), coloured by chain in cartoon display with the protein's motifs visible (figure C) and coloured by motif in cartoon mode (figure D). A-helices are in pink colour and β-sheets in orange colour.

Cerulenin is an irreversible inhibitor of the β-ketoacyl carrier synthase proteins I and II. It establishes a covalent bond with the Cysteine of the active site of the protein. It is not selective antibacterial since it has been found that it is a possible inhibitor of the condensation reaction, which is catalysed by the multifunctional fatty acid synthase type I that is found in mammals. It has been found that Cerulenin has anti-neoplastic activity from experiments in mice. Thiolactomycin, on the other hand, is a molecule that can inhibit, reversibly, the type II fatty acid synthases. Although, cannot do the same to the type I fatty acid synthases. Thiolactomycin shows activity against the Gram-negative anaerobes, from the periodontal disease, and from its property of inhibiting the mycolic acid Thiolactomycin is also an antimycobacterial agent. Finally Thiolactomycin can be used as an agent against malaria and trypanosomes, which makes it an ideal antimicrobial research tool.

The main aim of this project is to study the interactions that take place between the E.Coli KAS enzymes and the ligands. In this case the ligands will be Cerulenin and Thiolactomycin. Moreover, the next aim was the establishment of the Plasmodium KAS I/II and KAS III structures out of their sequences, by means of homology modelling. As soon as the model is established, docking of Cerulenin and Thiolactomycin would be the next step. The results from the docking would provide vital information about the nature of the KAS I/II and KAS III Plasmodium enzymes when compared to the existing structures of the E.Coli enzymes. Final objective would be to design a new compound-drug that would show higher affinity in binding the active sites of the Plasmodium enzymes and thus inhibit their activity.

2.3 Methods:

2.3.1 Active site studies of the enzymes:

The KAS I/II enzyme was extensively studied both with Thiolactomycin and Cerulenin docked in the enzyme's active site. In fact a variety of different structures were studied but the main focus was on the KAS I/II enzyme. The structures that were used in this study are summarised in the table below (table 2).

Table 1. The different structures of the KAS enzymes used in this study. All of the structures below were obtained from the Expasy-pdb database.

Structure ID	Residue	Inhibitor	Atoms	PDB	Res.
3-Ketoacyl-ACP Synthase I	406	No Inhibitor	---	1G5X	2.45
3-Ketoacyl-ACP Synthase I	500	Thiolactomycin	14	1FJ4	2.35
3-Ketoacyl-ACP Synthase I	500	Cerulenin	16	1FJ8	2.27
3-Ketoacyl-ACP Synthase I	901	Dodecanoic Acid	13	1F91	2.40
3-Ketoacyl-ACP Synthase III	350	No Inhibitor	---	1HN9	2.00
3-Ketoacyl-ACP Synthase III	350	Coa	48	1HND	1.60
3-Ketoacyl-ACP Synthase III	340	Malonyl-Coa	48	1HNJ	1.46
3-Ketoacyl ACP Synthase I	413	Dodecadienamide	16	1B3N	2.65
3-Ketoacyl-ACP Synthase II	416	No Inhibitor	---	1E5M	1.54
3-Ketoacyl-ACP Synthase I	901	Dodecanoic Acid	13	1EK4	1.85

Legend:

Structure ID : Name of the structure
Residue : Number of residues in structure per chain
Inhibitor : Name of complexed inhibitor in structure
Atoms : Number of Atoms in Inhibitor
PDB : The ID-code of the structure's pdb file in Expasy's PDB database
Res. : The resolution of the X-Ray crystallography of the structure

All the sequences of the various β-ketoacyl carrier protein synthases (from table 2) were imported into the LPC analysis software. LPC analysis software (Ligand - Protein Contacts analysis software) was used for the active site analysis studies. LPC automatically: 1. identifies the residues that make the active site of the protein; 2. calculates the solvent accessible surface of every atom; 3. determines the contacting residues and type of interaction they undergo (hydrophobic-hydrophobic, aromatic-aromatic, etc.); 4. indicates all putative hydrogen bonds. LPC also predicts changes in binding strength following chemical modification of the ligand. 5. calculates all the distances between the ligand and the protein's active site. The second software package that was used is the CSU program (Contacts of Structural Units). The CSU program calculates contacts of structural units such as helices, sheets, strands and residues. Both approaches are based on a detailed analysis of inter-atomic contacts and interface complementarity.

2.3.2 Superimpositions:

Superimposition of the structures was done with the SwissProt package by GlaxoSmithKline. The structures must be in pdb file format, which means that some files had to be edited to isolate the active sites and the subunits of the proteins. Both of the structures are uploaded into the software and then the option "magic fit" is selected from the tools menu. The superimposition of the structures can be improved by selecting the "iterative fit". Magic fit performs a crude superimposition of one structure onto the other one and iterative magic fit will improve the fit as much as possible. The two structures can be coloured, such that differences will then become very easy to spot and to identify. At this stage the RMS distance (RMSd), i.e. the RMS deviation between the backbone atoms of the two structures, was calculated.

2.3.3 Sequence Alignment:

The sequence alignment was done with a multiple alignment program called ClustalW (Higgins and Sharp, 1989). The sequences had to be in FASTA format to be able to be used as input for the program. The FASTA format is the sequence of the protein with. The whole process of the sequence alignment is broken down into four steps. Firstly, using rapid alignment methods the program calculates all pairwise similarity scores. The second step is the generation of a similarity matrix. Then the sequences are clustered according to the generated similarity matrix with the aid of an algorithm. Next step is the making of a cluster alignment using a consensus method and finally a multiple-progressive alignment is generated. The groups of the sequences are aligned according to their cluster branch order. Seven different structures of the KAS enzyme from different sources were aligned against each of the two sequences that were to be modelled. Those structures were all obtained from the Expasy-pdb database and their codes were: 1FJ4, 1FJ8, 1HN9, 1NHJ, 1EK4 and 1G5X. The first run involved all the above structures against the two sequences of the Plasmodium KAS I/II and KAS III enzymes. Then the structures with the best alignment identity percentages were aligned against the two sequences.

2.3.4 Threading

Another very important tool is threading. Threading (Linus Pauling *et al.* 1965) involves looking at the structure, in order to find residues that would be happy to occupy a certain place on the whole structure. For example, a leucine to aspartic acid mutation can not be characterised as disastrous or beneficial until the relative location of the mutation is found. A leucine to aspartic acid mutation near the protein's surface is beneficial whereas a leucine to aspartic acid mutation in the protein's core is disastrous. The SwissProt software has a built-in function to depict the threading energy of each residue. First one of the Plasmodium sequences and its template structure are imported into the SwissProt software. Then a structural alignment is done and a curve depicting how each residue "likes" its surrounding is displayed by calculating the mean force potential energy for this residue. If a residue is "happy", its energy is below zero, whereas an unhappy residue will have energy above the zero axis. Finally enabling the "Auto Color by Threading Energy" item from the menu will colour the residues according to their threading energy.

2.3.5 Homology modelling:

Homology modelling was done with Composer running on a SGI workstation. Homology modelling is the comparison of a protein sequence of unknown structure with that of a protein with known structure, in order to predict what the unknown structure should, ideally, look like. Each of the sequences of the Plasmodium KAS proteins were imported in COMPOSER and the sequence of the template protein were added in the database of the software. The template had to be in pdb format, and to be added in the existing database of COMPOSER. As soon as that was done, the program would search for homologues in its database and the imported sequence was found and selected. Then COMPOSER did a sequence alignment of the two sequences. Initially, there were 12 loops and the program only managed to fix only one of those, using its database. The database of COMPOSER at that time numbered 1800 structures. So, in order to solve the problem, a larger database with more than ten thousand structures was used. This fixed all eleven loops that were left and a model for each of the Plasmodium KAS protein sequences was built.

2.3.6 Energy minimization

Energy minimisations were done with HyperChem on a PC running Windows NT. Only Thiolactomycin was chosen to be energetically minimised due to time restrictions. The idea was to energetically minimize the conformation of Thiolactomycin to its nearest local minimum when suspended in an aqueous environment (i.e. just before it binds to the active site of the protein). That was done in order to observe how much conformational change Thiolactomycin undergoes, when it binds to the active site of the protein. Thiolactomycin was first minimised by MM+ and then by PM3 optimisations. In order to simulate the water environment a periodic box full with water molecules was used. The dimensions of the periodic box were 18,7 x 18,7 x 18,7 Å (default dimensions - chosen by the program). The figure below demonstrates how Thiolactomycin was optimised in the H2O environment.

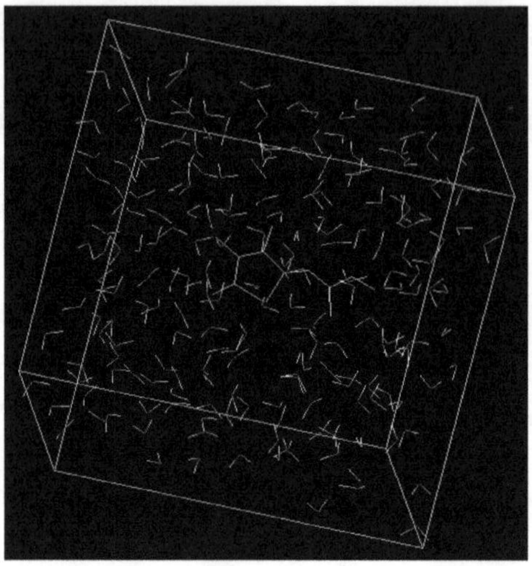

Figure 15. TLM optimisation in H_2O environment.

2.4 Results & Discussion:

2.4.1 Active site & enzyme studies:

The LPC analysis was performed on both 1FJ4 and 1FJ8, which is the same protein but with Thiolactomycin and Cerulenin docked respectively. The active sites were studied in terms of their distance from the ligand, their exposed contact accessible area and the type of bond that they establish. Four groups of interaction types are being generated by the LPC analysis: hydrophilic-hydrophilic contact (hydrogen bond), aromatic-aromatic contact, hydrophobic-hydrophobic contact, hydrophobic-hydrophilic contact.

As it can be seen from table 2 most varying interaction type among the subunits is the hydrophobic-hydrophilic contacts. Those bonds are not as strong as the h-bonds but are still very important to define the affinity of the ligand's binding to the active site of the protein. Hydrophobic interactions are the major driving force between non-polar groups and water that force the non-polar groups to aggregate together in order to minimise their surface area and contact with water molecules. It should be noted that hydrophobic/hydrophilic interactions belong to the dominant forces determining protein structure [10]. On the other hand hydrogen bonds are intermolecular forces between strongly charged dipoles located on different molecules or different parts of one large molecule. Although stronger than most other intermolecular forces, hydrogen bonds are much weaker than both the ionic and covalent bond. The hydrogen bond has only 5% or so of the strength of a covalent bond. However, when many hydrogen bonds can form between two molecules (or parts of the same molecule), the resulting union can be sufficiently strong as to be quite stable.

The differences in the interaction between the four subunits of the KAS enzyme when Thiolactomycin binds are shown in table 3. With Thiolactomycin there are only differences in the hydrophobic-hydrophilic contacts between the protein and the ligand. Because those differences though are not many and aren't particularly strong, one could even ignore them, although as it has already been mentioned, the hydrophobic-hydrophilic contacts are key contacts, which determine the protein structure.

On the other hand when Cerulenin binds to the KAS enzyme, there are more significant differences among the four subunits of the protein. In more detail, subunits A and B both establish a hydrogen bond between the residue Cys163 and the ligand, whereas subunits C and D don't. All the rest interactions between the ligand and the active site of the protein are exactly the same as far as Cys163 is concerned. Another case with varying hydrogen bond interactions is Thr300. This residue establishes a hydrogen bond with the ligand only in subunit C. The greatest difference among the active sites of the four subunits is the Val133 residue, which is only found in B subunit. The existence of Val133 in the active site of B subunit has an effect on the properties of this particular active site as it establishes a DC interaction with the ligand. One more difference between the active sites involves the non-establishment of DC bonds between the ligand and Gln113 in subunit D, the establishment of a DC bond between the ligand and Met138 in subunit C, the establishment of a DC bond between the ligand and Ala137 in subunit D and finally the establishment of a DC bond between the Cerulenin and Met197 only in subunits B and D of the protein.

Table 2. LPC analysis on the characteristics and the interactions that take place between the KAS I protein and Thiolactomycin (pdb entry: 1FJ4).

Residue	Dist				HB				Arom				Phob				DC			
163 CYS	3.2		3.2	3.3	-		-	-	-		-		-		-	-	+		-	-
206 ALA	5.5		5.4	5.6	-		-	-	-		-		+		+	+	-		-	-
229 PHE	3.4		3.5	3.5	-		-	-	-		-		+		+	+	-		-	-
265 ASP	5.2		5.1	4.4	-		-	-	-		-		-		-	-	+		+	+
268 ASP	3.8		3.7	3.7	-		-	-	-		-		-		-	-	+		+	+
270 VAL	3.5		3.5	3.6	+		+	+	-		-		-		-	-	+		+	+
271 ALA	3.3		3.2	3.3	-		-	-	-		-		-		-	-	+		+	+
272 PRO	3.9		3.8	3.8	-		-	-	-		-		+		+	+	+		+	+
298 HIS	2.8		3.0	2.9	+		+	+	-		-		+		+	+	-		-	-
300 THR	3.7		3.8	3.7	-		-	-	-		-		+		+	+	+		+	+
302 THR	4.0		4.0	3.9	+		+	+	-		-		+		+	+	+		+	+
305 GLY	4.0		4.0	4.0	+		+	+	-		-		-		-	-	-		-	+
333 HIS	2.6		2.7	2.7	+		+	+	-		-		+		+	+	+		-	-
390 PHE	3.3		3.3	3.2	-		-	-	-		-		+		+	+	+		-	-
391 GLY	3.5		3.6	3.5	-		-	-	-		-		-		-	-	+		+	+
392 PHE	3.7		3.6	3.9	-		-	-	-		-		+		+	+	-		-	-
393 GLY	3.8		3.9	3.9	-		-	-	-		-		-		-	-	+		+	+
394 GLY	3.3		3.3	3.3	-		-	-	-		-		-		-	-	+		+	+
396 ASN	5.0		5.0	4.8	-		-	-	-		-		-		-	-	+		+	+

Table 3. LPC analysis on the characteristics and the interactions that take place between the KAS I protein and Cerulenin (pdb entry: 1FJ8).

Residue	Dist				HB				Arom				Phob				DC			
106 GLY	3.2		3.1	3.3	-		-	-	-		-	-	-		-	-	+		+	+
107 GLY	3.5		3.6	3.6	-		-	-	-		-	-	-		-	-	+		+	+
110 PRO	4.3		4.3	4.4	-		-	-	-		-	-	+		+	+	-		-	-
162 ALA	3.3		3.3	3.3	-		-	-	-		-	-	+		+	+	-		-	-
163 CYS	1.9		1.9	1.9	+		-	-	-		-	-	-		-	-	+		+	+
191 GLU	4.7		4.9	4.8	-		-	-	-		-	-	-		-	-	+		+	+
197 MET	3.4		3.3	3.4	-		-	-	-		-	-	+		+	+	-		-	+
200 GLU	2.8		2.9	2.8	-		-	-	-		-	-	-		-	-	+		+	+
201 PHE	3.0		2.7	2.8	-		-	-	-		-	-	+		+	+	-		-	-
229 PHE	5.0		4.8	4.9	-		-	-	-		-	-	+		+	+	-		-	-
298 HIS	3.0		2.8	2.8	+		+	+	-		-	-	-		-	-	-		-	-
300 THR	4.0		3.8	3.9	-		+	-	-		-	-	-		-	-	+		+	+
333 HIS	2.8		2.6	2.7	+		+	+	-		-	-	+		+	+	+		+	+
335 LEU	4.2		4.3	4.2	-		-	-	-		-	-	+		+	+	-		-	-
390 PHE	3.7		3.8	3.7	+		+	+	-		-	-	-		-	-	+		+	+
391 GLY	2.9		2.9	2.8	-		-	-	-		-	-	-		-	-	-		-	-
392 PHE	2.7		2.7	2.7	+		+	+	-		-	-	+		+	+	-		-	-
113 GLN	3.5		2.9	4.3	-		-	-	-		-	-	+		+	+	+		+	-
133 VAL	X		X	X	X		X	X	X		X	X	X		X	X	X		X	X
134 VAL	3.3		3.4	3.5	-		-	-	-		-	-	+		+	+	+		+	+
137 ALA	3.3		3.3	3.4	-		-	-	-		-	-	+		+	+	-		-	+
138 MET	3.4		3.4	3.6	-		-	-	-		-	-	+		+	+	-		+	-

Legend:

Colours: Red = subunit A, Green = subunit B, Blue = subunit C, Gray = subunit D.
Dist: nearest distance (Å) between atoms of the ligand and the residue
HB: hydrophilic-hydrophilic contact (hydrogen bond)
Arom: aromatic-aromatic contact
Phob: hydrophobic-hydrophobic contact
DC: hydrophobic-hydrophilic contact (destabilizing contact)
+/-: indicates presence/absence of a specific contacts
Underlined: indicates residues contacting ligand by their side chain (including CA atoms)

The active site of the 1FJ4, which was isolated using the LPC analysis software, is depicted in the following figures. 1FJ4 is a complex of the KAS protein with Thiolactomycin. The pdb file was edited and only the Subunit A was retained. Then basing on the results of the LPC analysis, only the residues of the active site and the ligand were retained. Following the same procedures the active site of KAS I without any ligand docked was also isolated (pdb entry: 1G5X). Figures 16a and 16b below, show the active site of the KAS I enzyme, with and without the ligand docked respectively (in 16a Thiolactomycin is in yellow). In figures 16c and 16d the two active sites have been superimposed. In figure 16c the active site without the ligand is in blue and in 16d is in red. It is obvious that the conformation of the active site changes significantly when the ligand binds to it. The RMDd value of the two active site superimposition was found to be 0.67 A, with the 19 residues of each of the two active sites involved in the calculation.

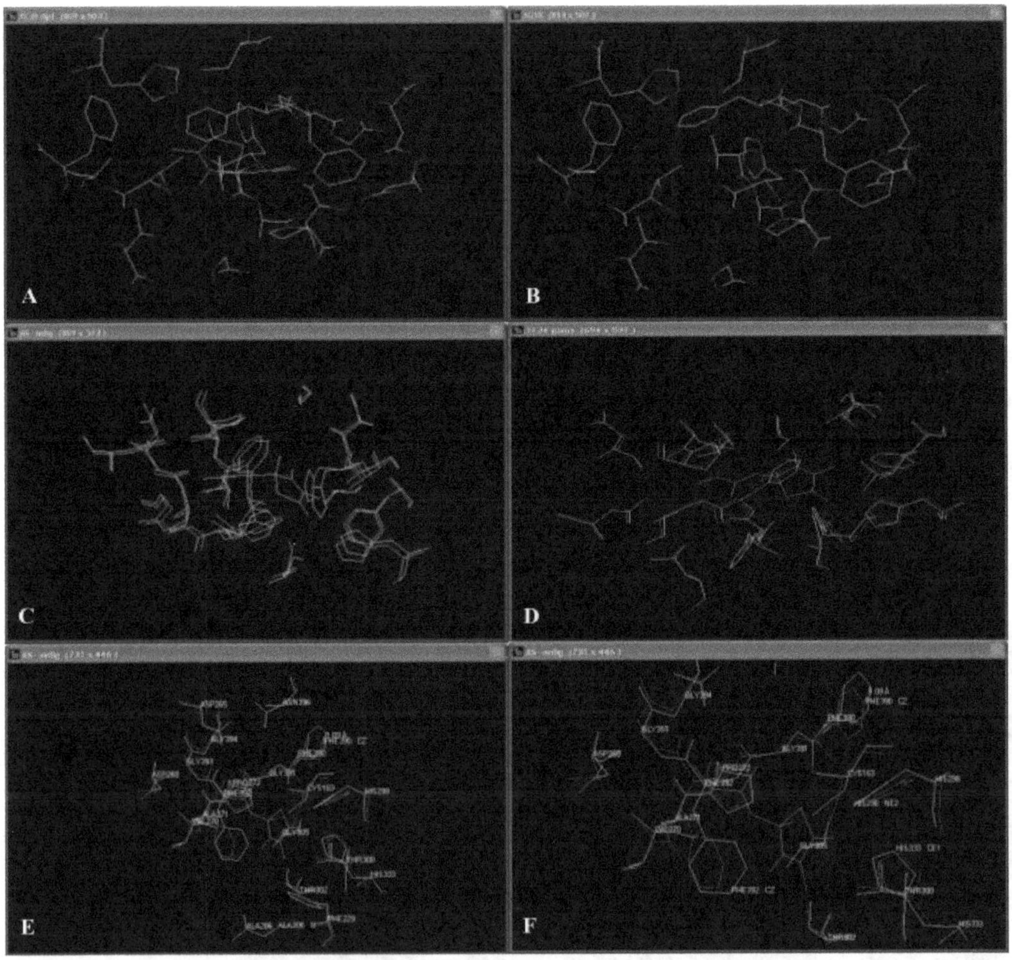

Figure 16. Active site upon LPC analysis

From the LPC analysis it was found that the contact surface area (calculated in Å^2) between the ligand and the residue is directly proportional to the degree of the residue's movement when the ligand binds to the active site. In the following table the contact surface area (Å^2) and the movement distance (Å) of some residues have been summarised.

Table 4. Contact surface area (Å^2) and movement distance (Å) of some residues in the KAS I protein's active site.

Residues	Surface (Å^2)	Distance (Å)
Phe390	46.4	1.09
Phe392	30.1	1.21
His298	21.7	1.68
Ala206	1.8	0.13
Asp265	0.4	0.21

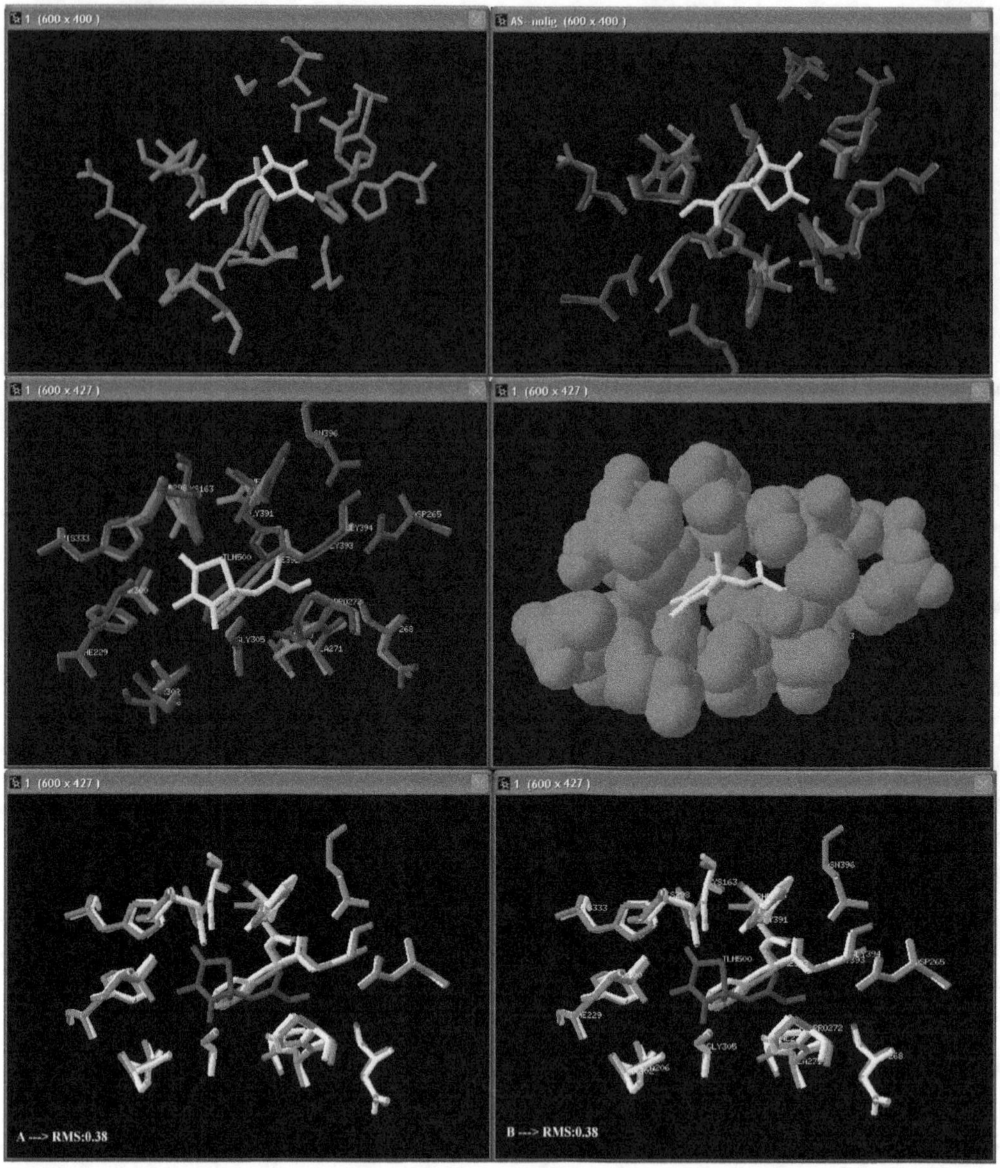

Figure 17. The interactions that Thiolactomycin establishes with the active site of the KAS I enzyme (data from LPC analysis of the 1FJ4 pdb file).

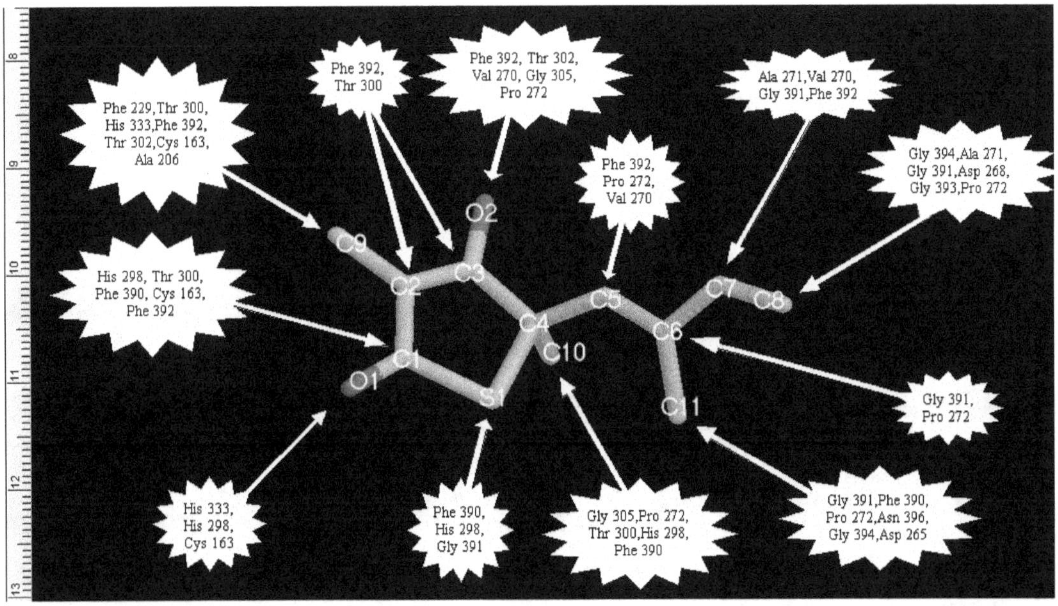

Figure 18. The interactions that Cerulenin establishes with the active site of the KAS I enzyme (data from LPC analysis of the 1FJ8 pdb file). Note that Cerulenin only interacts with Val134 (in bold) in subunit B.

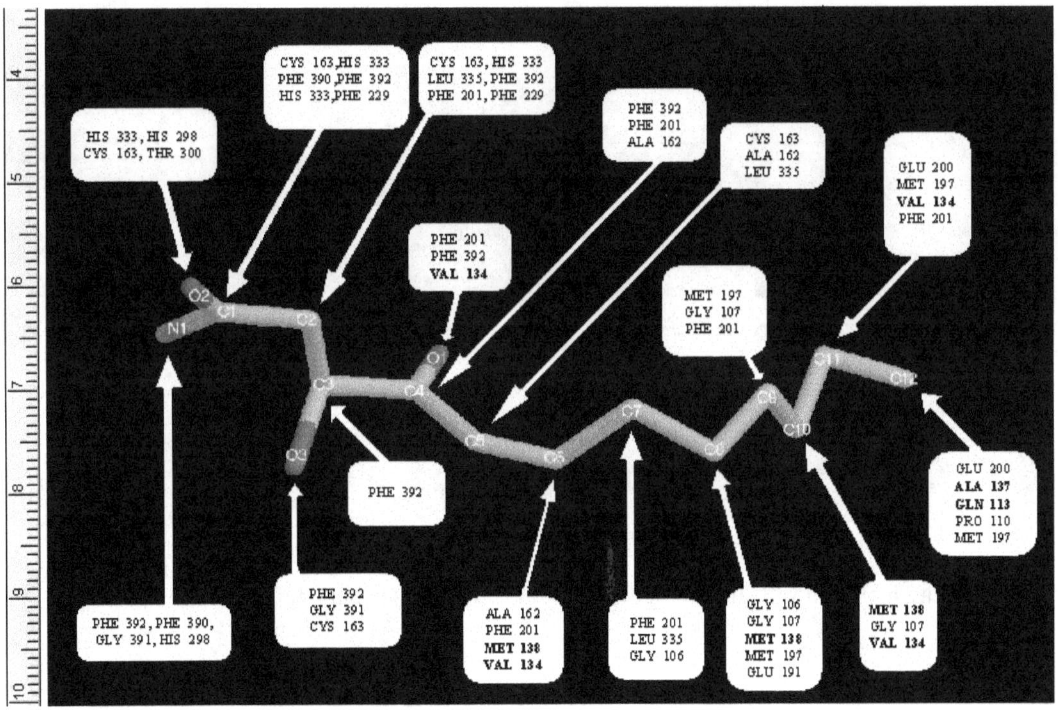

Figure 19. Interaction between Thiolactomycin and the active site of KAS I as shown using LIGPLOT.

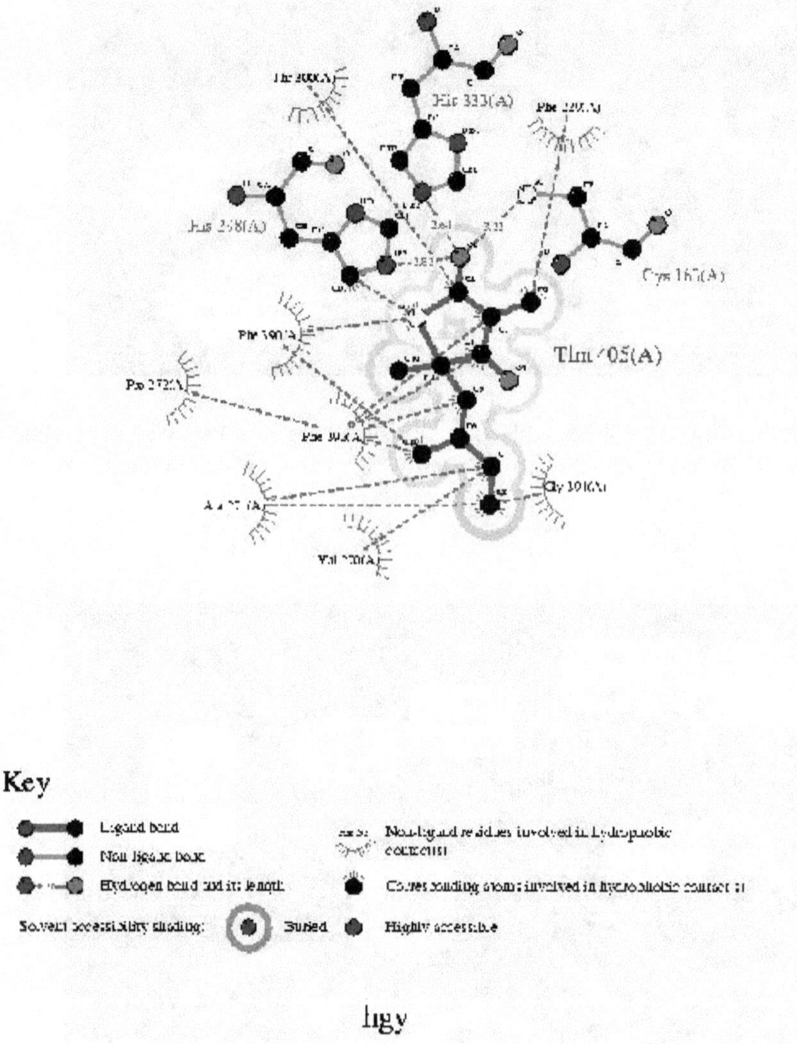

Figure 20. Interaction between Thiolactomycin and the active site of KAS I as shown using LIGPLOT.

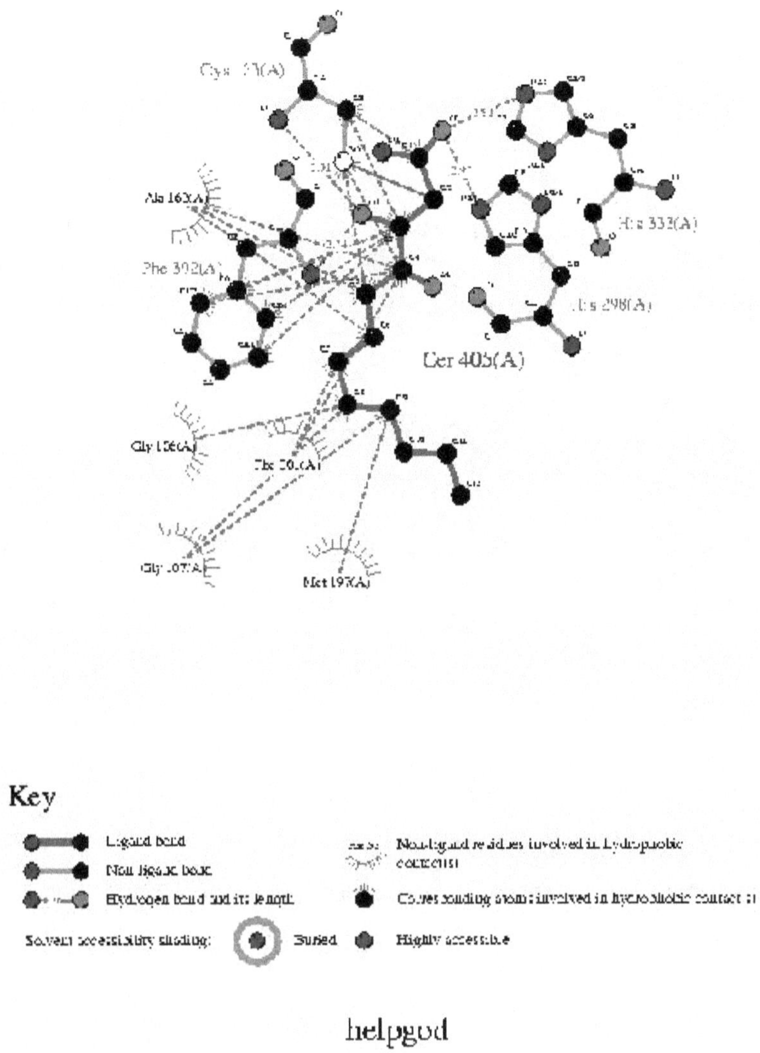

Figure 21. Interaction between Thiolactomycin and the active site of KAS I as shown using LIGPLOT.

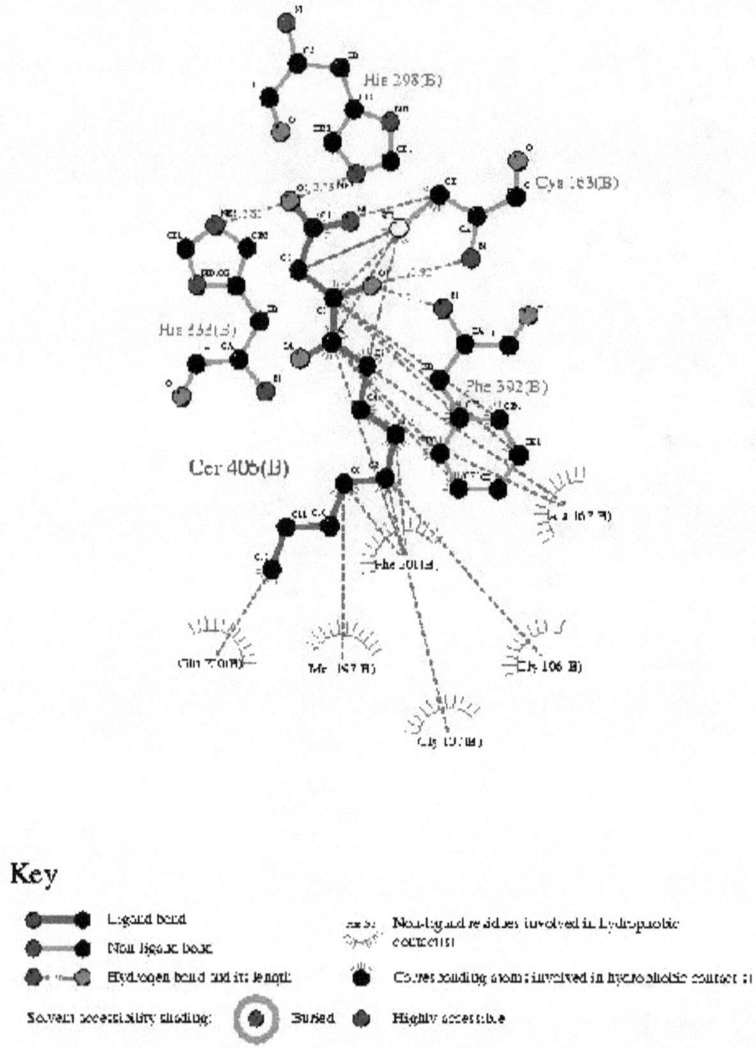

jkj

Figure 22. Different views of the superimposition of the four subunits of the KAS I enzyme with Cerulenin docked (pdb entry: 1FJ8).

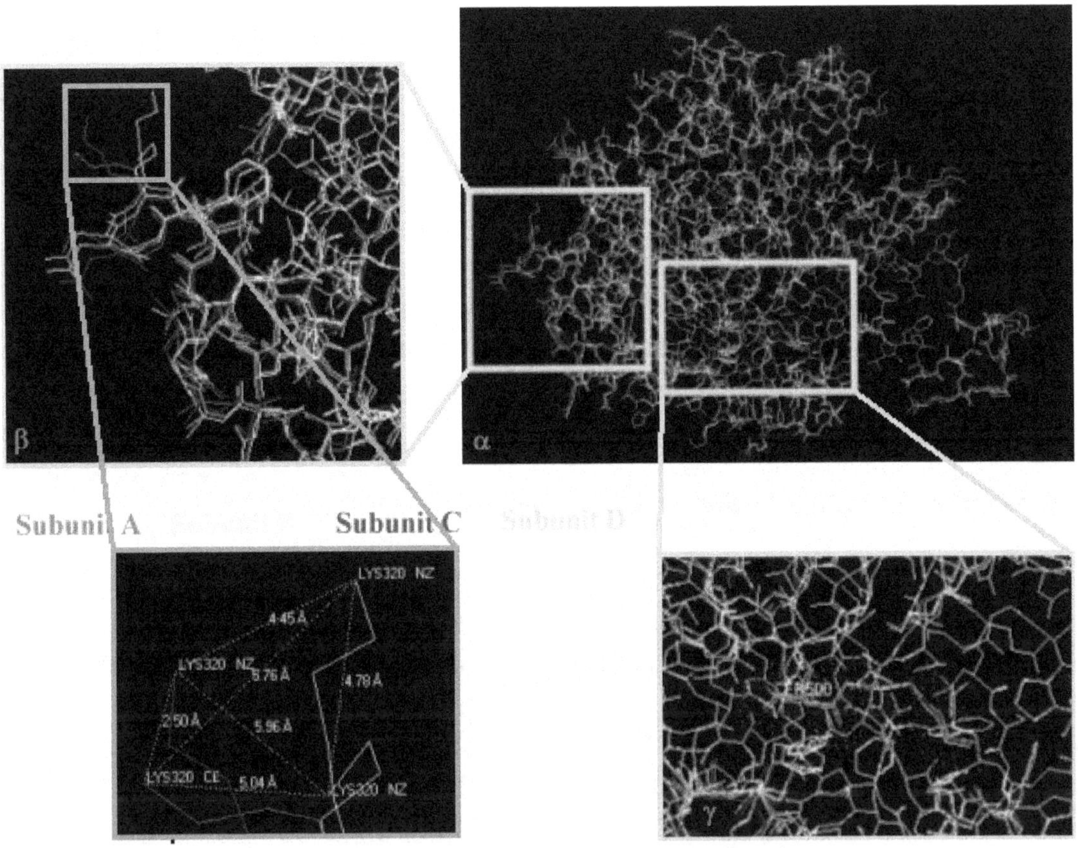

Figure 23. The active site region

The LPC analysis revealed that there are a lot of differences between the four active sites of the KAS I enzyme when Cerulenin binds to it. In order to investigate those conformational differences between the four subunits of the KAS I protein (with Cerulenin in the active site), the four subunits were superimposed. Figure 23a shows the four superimposed subunits and below there is a subunit-colour reference too. Then in 23b a part of the superimposition near the outside of the four subunits is magnified, whereas in 23c the superimposed area around the active sites of the subunits is magnified. It is very clear that the superimposition in the active-site area of the subunits has been done smoothly, as there are not big movements and conformational changes. On the other hand on the outside of the protein a lot differences between the four subunits can be observed. In figure 23d a part of 23b has been further zoomed in and the distances between same residues on different subunits have been measured. Those are huge differences as they can be as big as 5.76 A. One would assume that those differences are not very significant to the function of the protein as they are far away from the active site, but that is not true. According to the work done by Olsen, Kadziola et al [ref], on KAS I it has been found that there is a path that the ligand has to follow to reach the active site of the protein. As a result the conformation of the outside of the protein might have an effect on the easiness or difficulty that the ligand will have to cope with in order to reach the active site. Thus the conformation of the outside of the subunit could affect the affinity of the subunit for binding the ligand as a whole. The figure (figure 24) below shows part of the work done by Olsen, Kadziola et al, emphasising in the entrance of the active site of the KAS I subunit.

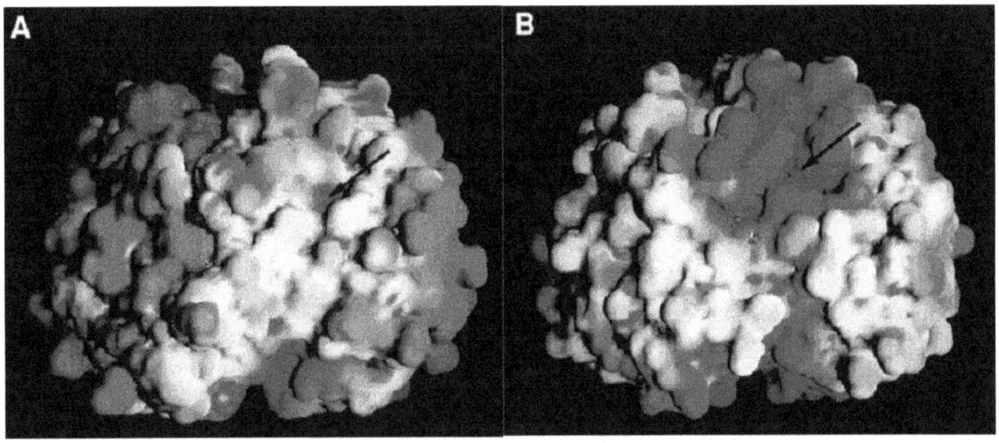

Fig. 24 A and B: Accessible surface map representation of electrostatic potential distribution on the KAS I and KAS II dimer, respectively. Positive and negative electrostatic potentials are colored blue and red, respectively. Notice that the overall distribution of charges is the same: Positive potential around the active site flanked by strong negative potential. An arrow points to the active site entrances. (figures courtesy of Olsen et al.).

The above idea among the four subunits of the same protein (KAS I) can be further extended on different protein of the same family. So, conformational changes between KAS I and KAS II proteins can be responsible for ligand selectivity and binding affinity.

Given the availability of C_{10} and C_{12} fatty acid–KAS I complexes (J. G. Olsen, A. Kadziola, P. von Wettstein-Knowles, M. Siggaard-Andersen and S. Larsen, unpublished work) and the cerulenin–KAS II complex [27], the substrate-binding cavities of the two enzymes can be compared. The architecture of the two enzymes, however, is identical in the pertinent region adjacent to the active-site cysteine where one would predict KAS I but not KAS II to be able to harbour a *cis* double bond in the δ3 $C_{10:1}$ substrate. This conclusion is based on modelling a Z-bond to resemble a *cis* bond in place of the E-bond found in both the C_{10} and C_{12} substrates without encountering steric hindrance (J. G. Olsen, A. Kadziola, P. von Wettstein-Knowles, M. Siggaard-Andersen and S. Larsen, unpublished work). The possibility thus arises that specificity results from substrate selection before its entrance into the binding pocket. With respect to the length of the substrate that can be elongated by KAS I and KAS II, KAS I accepts C_{16} acyl chains from acyl-ACP as readily as C_{14} chains [10]. Under glycerol-starvation conditions *E. coli* synthesizes prominent amounts of $C_{20:1}$, C_{18} and C_{20} as well as traces of C_{22}, which occur as free acids [28]. This implies that KAS I and/or KAS II can carry out as many as three additional elongation steps, which makes the reluctance of either isoenzyme to extend C_{16} *in vitro* [9] a puzzle. The KAS II binding pocket is more hydrophobic than that of KAS I, which ends at the dimer interface with charged/polar residues (Glu-200, Gln-113). By comparison the KAS II binding pocket does not have an obvious end where the two pockets meet on the interface [12]. Obviously more work is required if we are to understand the subtle nature of the structural differences giving rise to the characterized substrate specificities in the KAS family.

2.4.2 Multiple Sequence Alignments

<u>Sequence Alignment 1:</u>

Sequence 1: kas3 371 amino acids
Sequence 2: kas1/2 443 amino acids
Sequence 3: 1FJ4 406 amino acids
Sequence 4: 1E5M 416 amino acids
Sequence 5: 1DD8 406 amino acids
Sequence 6: 1EBL 317 amino acids
Sequence 7: 1HN9 317 amino acids
Sequence 8: 1HZP 335 amino acids
Sequence 9: 1KAS 412 amino acids

<u>Alignment Percentage Identities:</u>

Sequences (1:2) Aligned. Score: 7 percent
Sequences (1:3) Aligned. Score: 5 percent
Sequences (1:4) Aligned. Score: 2 percent
Sequences (1:5) Aligned. Score: 5 percent
Sequences (1:6) Aligned. Score: 35 percent
<u>Sequences (1:7) Aligned. Score: 35</u> percent
Sequences (1:8) Aligned. Score: 29 percent
Sequences (1:9) Aligned. Score: 11 percent
Sequences (2:3) Aligned. Score: 31 percent
<u>Sequences (2:4) Aligned. Score: 45</u> percent
Sequences (2:5) Aligned. Score: 31 percent
Sequences (2:6) Aligned. Score: 5 percent
Sequences (2:7) Aligned. Score: 9 percent
Sequences (2:8) Aligned. Score: 11 percent
Sequences (2:9) Aligned. Score: 37 percent
Sequences (3:4) Aligned. Score: 33 percent
Sequences (3:5) Aligned. Score: 99 percent
Sequences (3:6) Aligned. Score: 13 percent
Sequences (3:7) Aligned. Score: 12 percent
Sequences (3:8) Aligned. Score: 9 percent
Sequences (3:9) Aligned. Score: 35 percent
Sequences (4:5) Aligned. Score: 34 percent
Sequences (4:6) Aligned. Score: 10 percent
Sequences (4:7) Aligned. Score: 10 percent
Sequences (4:8) Aligned. Score: 9 percent
Sequences (4:9) Aligned. Score: 50 percent
Sequences (5:6) Aligned. Score: 13 percent
Sequences (5:7) Aligned. Score: 12 percent
Sequences (5:8) Aligned. Score: 9 percent
Sequences (5:9) Aligned. Score: 35 percent
Sequences (6:7) Aligned. Score: 97 percent
Sequences (6:8) Aligned. Score: 35 percent
Sequences (6:9) Aligned. Score: 9 percent
Sequences (7:8) Aligned. Score: 35 percent
Sequences (7:9) Aligned. Score: 12 percent
Sequences (8:9) Aligned. Score: 11 percent

<u>Alignment Scores:</u>

There are 8 groups

Group 1: Sequences: 2 Score:6802
Group 2: Sequences: 3 Score:6430
Group 3: Sequences: 2 Score:8764
Group 4: Sequences: 5 Score:4530
Group 5: Sequences: 2 Score:6714
Group 6: Sequences: 3 Score:4602
Group 7: Sequences: 4 Score:3647
Group 8: Sequences: 9 Score:2185
Alignment Score 12287

Multiple sequence alignment:

```
1E5M            --------------------MANLEKKRVVVTGLGAITPIGNTLQDYWQGLMEGRNGIGP 40
1KAS            ------------------------SKRRVVVTGLGMLSPVGNTVESTWKALLAGQSGISL 36
kas1/2          MIKVIICYTFIFIYILIFLILLFFITSRVVCTGVGVVTGLGIGIEHFWNNIINGYTSIDK 60
1FJ4            -------------------------MKRAVITGLGIVSSIGNNQQEVLASLREGRSGITF 35
1DD8            -------------------------MKRVVITGLGIVSSIGNNQQEVLASLREGRSGITF 35
1EBL            ------------------------------------------------------------
1HN9            ------------------------------------------------------------
kas3            ------------------MFLYFITYLCIFHNNIYSVELIKNNKYNFINNVHNIKYRTKI 42
1HZP            --------------------------------------------------------MTEI 4
                                               ..

1E5M            ITRFDASDQACRFGGEVK--DFDATQFLD-RKEAKRMDRFCHFAVCASQQAINDAKLVIN 97
1KAS            IDHFDTSAYATKFAGLVK--DFNCEDIIS-RKEQRKMDAFIQYGIVAGVQAMQDSGLEIT 93
kas1/2          ITKFDITGMSCGIGSEIKKSDFNPSDYYTNKKDVNRNDDCTHYAVAATRLALDDAKLNLE 120
1FJ4            SQELKDSGMRSHVWGNVK---LDTTGLID-RKVVRFMSDASIYAFLSMEQAIADAGLSPE 91
1DD8            SQELKDSGMRSHVWGNVK---LDTTGLID-RKVVRFMSDASIYAFLSMEQAIADAGLSPE 91
1EBL            --XY----TKIIGTGSYL------------PEQVRTNADLEKXVDTSDEWIVTRTGIRER 42
1HN9            --MY----TKIIGTGSYL------------PEQVRTNADLEKMVDTSDEWIVTRTGIRER 42
kas3            RAIYGKTGGKIIGHGHSY------------PSTEIYNDELKKYVDTNDEWIRTRTGIKKR 90
1HZP            ATTSGARSVGLLSVGAYR------------PERVVTNDEICQHIDSSDEWIYTRTGIKTR 52
                              .                .                      : :

1E5M            ELNAD-EIGVLIGTGIGGLKVLEDQQTILLD-KGPSRCSPFMIPMMIANMASGLTAINLG 155
1KAS            EENAT-RIGAAIGSGIGGLGLIEENHTSLMN-GGPRKISPFFVPSTIVNMVAGHLTIMYG 151
kas1/2          KLDKD-KTGTIIGSGIGGLRFLEKEMKTMYE-KGHKRITPYLIPAMIANTPSGYVSIENN 178
1FJ4            AYQNNPRVGLIAGSGGGSPRFQVFGADAMRGPRGLKAVGPYVVTKAMASGVSACLATPFK 151
1DD8            AYQNNPRVGLIAGSGGGSPRFQVFGADAMRGPRGLKAVGPYVVTKAMASGVSACLATPFK 151
1EBL            HIAAPNETVSTXGFEAATRAIEX---AGIEKDQIGLIVVATTSATHAFPSAACQIQSXLG 99
1HN9            HIAAPNETVSTMGFEAATRAIEM---AGIEKDQIGLIVVATTSATHAFPSAACQIQSMLG 99
kas3            RILKRDENISMLQIDSATQALET---SCLKPSDIDMVINASSTPQNLFG-DANNISNKIG 146
1HZP            RFAADDESAASMATEACRRALSN---AGLSAADIDGVIVTTNTHFLQTPPAAPMVAASLG 109
                      .             .       :          .           :

1E5M            AKG-PNNCTVTACAAGSNAIGDAFRLVQNGYAKAMICGGTEAAITPLSYAGFASARALSF 214
1KAS            LRG-PSISIATACTSGVHNIGHAARIIAYGDADVMVAGGAEKASTPLGVGGFGAARALST 210
kas1/2          IRG-ISLGMLSACATSGNTIGEAYRYIKYKEYDVMICGGTEASITPISFAGFNSLKALCT 237
1FJ4            IHG-VNYSISSACATSAHCIGNAVEQIQLGKQDIVFAGGGEELCWEMACE-FDAMGALST 209
1DD8            IHG-VNYSISSACATSAHCIGNAVEQIQLGKQDIVFAGGGEELCWEMACE-FDAMGALST 209
1EBL            IKGCPAFDVAAACAGFTYALSVADQYVKSGAVKYALVVGSDVLARTCDPT---------- 149
1HN9            IKGCPAFDVAAACAGFTYALSVADQYVKSGAVKYALVVGSDVLARTCDPT---------- 149
kas3            CKNSVNMDLTAACTGFIFAFVTAYNFLNR--YKNILIVGSDALSNFVDWR---------- 194
1HZP            AKGILGFDLSAGCAGFGYALGAAADMIRGGGAATMLVVGTEKLSPTIDMY---------- 159
                 :.       :.*:     :  *   :        .  * :

1E5M            -RNDDPLHASRPFDKDRDGFVMGEGSGILILEELESALARGAKIYGEMVGYAMTCDAYHI 273
1KAS            -RNDNPQAASRPWDKERDGFVLGDGAGMLVLEEYEHAKKRGAKIYAELVGFGMSSDAYHM 269
kas1/2          GYNDNPKKGCRPFDLKRSGFVMGEGSGILILESYEHAIKRNAPIYGEIISYSSECDAYHI 297
1FJ4            KYNDTPEKASRTYDAHRDGFVIAGGGGMVVVEELEHALARGAHIYAEIVGYGATSDGADM 269
1DD8            KYNDTPEKASRTYDAHRDGFVIAGGGGMVVVEELEHALARGAHIYAEIVGYGATSDGADM 269
1EBL            --------------DRGTIIIFGDGAGAAVLAASEEPG---------IISTHLHADGSYG 186
1HN9            --------------DRGTIIIFGDGAGAAVLAASEEPG---------IISTHLHADGSYG 186
kas3            --------------DRNTCVLFGDAAGAVVLQRTEEKEEN------KIFNYYLGSDSELN 234
1HZP            --------------DRGNCFIFADGAAAVVVGETPFQG---------IGPTVAGSDGEQA 196
                                   .::. ...  ::               .:      .*.
```

```
1E5M       TAPVPDGRGATRAIAWALKDSGLKPEMVSYINAHGTSTPANDVTETRAIKQALGNHAYNI  333
1KAS       TSPPENGAGAALAMANALRDAGIEASQIGYVNAHGTSTPAGDKAEAQAVKTIFGEAASRV  329
kas1/2     TAPEPNGKGLTNSIHKALKNANININDVKYINAHGTSTNLNDKIETKVFKNVFKDHAYKL  357
1FJ4       VAP--SGEGAVRCMKMAMHG---VDTPIDYLNSHGTSTPVGDVKELAAIREVFGDKSP--  322
1DD8       VAP--SGEGAVRCMKMAMHG---VDTPIDYLNSHGTSTPVGDVKELAAIREVFGDKSP--  322
1EBL       ELLTLP------NADRVNPENSIHLTXAGNEVFKVAVTELAHIVDETLAANNNDRSQLDW  240
1HN9       ELLTLP------NADRVNPENSIHLTMAGNEVFKVAVTELAHIVDETLAANNLDRSQLDW  240
kas3       DLLTINFDHDKYNLDKPNVNKYGKLYMNGKEVFKYTISNIPKILKKAIQHSNINIEDINY  294
1HZP       DAIRQDIDWITFAQNPSGPRP--FVRLEGPAVFRWAAFKMGDVGRRAMDAAGVRPDQIDV  254
                                               : :    .

1E5M       AVSSTKSMTGHLLGGSGGIEAVATVMAIAEDKVPPTINLENPDPECD-LDYVPGQ-SRAL  391
1KAS       LVSSTKSMTGHLLGAAGAVESIYSILALRDQAVPPTINLDNPDEGCD-LDFVPHE-ARQV  387
kas1/2     YISSTKSMTGHCIGAAGAIESIVCLKTMQTNIIPPTINYEYKDPDCD-LNYTPNKYIHAK  416
1FJ4       AISATKAMTGHSLGAAGVQEAIYSLLMLEHGFIAPSINIEELDEQAAGLNIVTET---TD  379
1DD8       AISATKAMTGHSLGAAGVQEAIYSLLMLEHGFIAPSINIEELDEQAAGLNIVTET---TD  379
1EBL       LVPHQANLR---IISATAKKLGXSXDNVVVT-LDRHGNTSAASVPCALDEAVRDG---RI  293
1HN9       LVPHQANLR---IISATAKKLGMSMDNVVVT-LDRHGNTSAASVPCALDEAVRDG---RI  293
kas3       FIFHQANIR---IIETVAKNLNIPMSKVLVN-LDEYANTSAASIPLCFSENIKNG---KI  347
1HZP       FVPHQANSR---INELLVKNLQLRPDAVVANDIEHTGNTSAASIPLAMAELLTTG---AA  308
            :         :      :       :    :    * .  .        :

1E5M       I-VDVALSNSFGFGGHNVTLAFKKYQ-  416
1KAS       SGMEYTLCNSFGFGGTNGSLIFKKI--  412
kas1/2     ENIDISLNTNLGFGGHNTALLFKKIVK  443
1FJ4       RELTTVMSNSFGFGGTNATLVMRKLKD  406
1DD8       RELTTVMSNSFGFGGTNATLVMRKLKD  406
1EBL       KPGQLVLLEAFGGGFTWGSALVRF---  317
1HN9       KPGQLVLLEAFGGGFTWGSALVRF---  317
kas3       KTNDIICMCGFGAGMSYGCVILKY---  371
1HZP       KPGDLALLIGYGAGLSYAAQVVRMPKG  335
                      * *         .:
```

Sequence Alignment 2:

Sequence 1: kas12 : 443 amino acids

Sequence 2: 1E5M_A : 416 amino acids

Sequences (1:2) Aligned. Score : 45 percent

Alignment Score : 1193

Multiple sequence alignment:

```
kas12      MIKVIICYTFIFIYILIFLILLFFITSRVVCTGVGVVTGLGIGIEHFWNNIINGYTSIDK 60
1E5M_A     --------------------MANLEKKRVVVTGLGAITPIGNTLQDYWQGIMEGRNGIGP 40
            .   . :               :  : ..*** **:*.:* :*  ::.:*:.::* ..*.

kas12      ITKFDITGMSCGIGSEIKKSDFNPSDYYTNKKDVNRNDDCTHYAVAATRLALDDAKLNLE 120
1E5M_A     ITRFDASDQACRFGGEVK--DFDATQFLD-RKEAKRMDRFCHFAVCASQQAINDAKLVIN 97
           **:** :. :* :*.*:*.:**:.:::  .:*:.:**   *:**.*:: *:**** ::

kas12      KLDKDKTGTIIGSGIGGLRFLEKEMKTMYEKGHKRITPYLIPAMIANTPSGYVSIENNIR 180
1E5M_A     ELNADEIGVLIGTGIGGLKVLEDQQTILLDKGPSRCSPFMIPMMIANMASGLTAINLGAK 157
           :*: *: *.:**:*****:.**.: . : :** .* :*::** **** .** .:*: . :

kas12      GISLGMLSACATSGNTIGEAYRYIKYKEYDVMICGGTEASITPISFAGFNSLKALCTGYN 240
1E5M_A     GPNNCTVTACAAGSNAIGDAFRLVQNGYAKAMICGGTEAAITPLSYAGFASARALSFR-N 216
           *  .    ::***:..*:**:*:* ::      ..********:***:*:*** * :**.   *

kas12      DNPKKGCRPFDLKRSGFVMGEGSGILILESYEHAIKRNAPIYGEIISYSSECDAYHITAP 300
1E5M_A     DDPLHASRPFDKDRDGFVMGEGSGILILEELESALARGAKIYGEMVGYAMTCDAYHITAP 276
           *:* :..**** .*.*************. * *: *.* ****::.*:  ********
```

```
kas12           EPNGKGLTNSIHKALKNANININDVKYINAHGTSTNLNDKIETKVFKNVFKDHAYKLYIS 360
1E5M_A          VPDGRGATRAIAWALKDSGLKPEMVSYINAHGTSTPANDVTETRAIKQALGNHAYNIAVS 336
                 *:*:* *.:*  ***::.:: : *.*********  **  **:.:*:.: :***:: :*

kas12           STKSMTGHCIGAAGAIESIVCLKTMQTNIIPPTINYEYKDPDCDLNYTPNKYIHAKENID 420
1E5M_A          STKSMTGHLLGGSGGIEAVATVMAIAEDKVPPTINLENPDPECDLDYVPGQ--SRALIVD 394
                ******** :*.:*.**::. : ::  : :***** *  **:***:*.*.:         :*

kas12           ISLNTNLGFGGHNTALLFKKIVK 443
1E5M_A          VALSNSFGFGGHNVTLAFKKYQ- 416
                ::*...:******.:* ***   .
```

Sequence Alignment 3:

Sequence 1: kas3_sequence	:	371 amino acids
Sequence 2: 1HN9_A	:	317 amino acids
Sequences (1:2) Aligned. Score	:	35 percent
Alignment Score	:	645

Multiple sequence alignment:

```
kas3_sequence     MFLYFITYLCIFHNNIYSVELIKNNKYNFINNVHNIKYRTKIRAIYGKTGGKIIGHGHSY 60
1HN9_A            ------------------------------------------MY----TKIIGTGSYL 12
                      :  .   ..  : .  .... .   ..   . .   :.  ::*..:. **** *

kas3_sequence     PSTEIYNDELKKYVDTNDEWIRTRTGIKKRRILKRDENISMLQIDSATQALETSCLKPSD 120
1HN9_A            PEQVRTNADLEKMVDTSDEWIVTRTGIRERHIAAPNETVSTMGFEAATRAIEMAGIEKDQ 72
                  *.     * :*:* ***.**** *****::*:*    :*.:* : :::**:*:* : :: .:

kas3_sequence     IDMVINASSTPQNLFG-DANNISNKIGCKNSVNMDLTAACTGFIFAFVTAYNFLNR--YK 177
1HN9_A            IGLIVVATTSATHAFPSAACQIQSMLGIKGCPAFDVAAACAGFTYALSVADQYVKSGAVK 132
                  *.::: *:::. : *    * :*.. :* *..  :*::***:** :*: .* ::::     *

kas3_sequence     NILIVGSDALSNFVDWRDRNTCVLFGDAAGAVVLQRTEEKEENKIFNYYLGSDSELNDLL 237
1HN9_A            YALVVGSDVLARTCDPTDRGTIIIFGDGAGAAVLAASEEPG---IISTHLHADGSYGELL 189
                    *:****.*:.  *  **.* ::***.***.**  :**  ...*:. :* :*.. .:**

kas3_sequence     TINFDHDKYNLDKPNVNKYGKLYMNGKEVFKYTISNIPKILKKAIQHSNINIEDINYFIF 297
1HN9_A            TLP------NADRVNPENSIHLTMAGNEVFKVAVTELAHIVDETLAANNLDRSQLDWLVP 243
                  *:  . .. * *: * ::  :* * *:**** :::::.:*:.:::  .*:: .::::::

kas3_sequence     HQANIRIIETVAKNLNIPMSKVLVNLDEYANTSAASIPLCFSENIKNGKIKTNDIICMCG 357
1HN9_A            HQANLRIISATAKKLGMSMDNVVVTLDRHGNTSAASVPCALDEAVRDGRIKPGQLVLLEA 303
                  ****:***.:.**:*.:.*.:*:*.**.:.******:*  .:.* :::*:**..::: : .

kas3_sequence     FGAGMSYGCVILKY 371
1HN9_A            FGGGFTWGSALVRF 317
                  **.*:::*..::::
```

Table 5. In an alignment if either ALN or GCG is selected in the output format then colouring of residues takes place according to the following physiochemical criteria:

Colour	Meaning
RED	Small (small+ hydrophobic (incl.aromatic -Y))
BLUE	Acidic
MAGENTA	Basic
GREEN	Hydroxyl + Amine + Basic - Q
Gray	Other

Table 6. In each alignment the following symbols are displayed, denoting the degree of conservation observed in each column:

Sign	Meaning
*	residues or nucleotides in that column are identical in all sequences in the alignment.
:	conserved substitutions have been observed, according to the COLOUR table above.
.	semi-conserved substitutions are observed.

The above sequence alignment study was performed using ClustaW, which is an online structural alignment tool [ref], in order for the template proteins to be chosen. Sybyl's subroutine Composer is also capable of searching for homologs that are similar to the unknown sequence that is imported but there is the drawback that COMPOSER is only capable of searching its own limited database and to pick up the homologs with the highest similarity score (percentage identity) to the imported sequence. The Composer software that was installed on our SGI machines had a database of approximately 1800 structures and the highest percentage identity that it was able to find was 16,7 percent, which is really low and could not be used as the basis of the building of a reliable model. The need for a larger database lead to expasy's web-based (online) database. There it is possible to search for "structural neighbours" of a known structure. So, a search was performed for structural neighbours of 1FJ4 and seven structures were isolated. Then using ClustalW the above sequence alignment was done and the percentage identity of those seven structures against the two unknown structures was found. The best percentage identity for KAS I/II was 45 percent and that was with 1E5M. Same way, the best percentage identity for KAS III was 35 percent and that was with both 1EBL and 1HN9. The similarity of 1EBL and 1HN9 was then tested and it was found to be 97 percent. So, the non-identical part of between the 1EBL and 1HN9 (3%) must be in the non-identical part (65%) between the unknown sequence and 1EBL or 1HN9. As a result, both 1EBL and 1HN9 could be used as the template to be used in the homology modelling. Eventually 1HN9 was chosen because the literature that came with the pdb file was a lot more detailed than the literature that came with 1EBL.pdb file. Importing those two pdb files into the COMPOSER's database produced the same results as with ClustalW. Both percentages were above the empirical minimum limit of 30% identity so, homology modelling was achievable.

2.4.3 Threading

The sequences of the KAS I and KAS II Plasmodium enzymes along with there templates were then incorporated into the SwissProt SPDBV Viewer (version 3.7). Then another sequence alignment was done. Using this sequence alignment a curve is generated, showing how much each residue enjoys or likes its surroundings. When a residue is "happy" in terms of its surrounding then its energy is low, below zero. On the other hand, if a residue is unhappy its energy is above zero, depending on how "unhappy" it is. This is the mean force potential energy. The smoothing factor was set to 1, which made the program to average the energy of a particular residue plus the energy of one flanking residue on each side. Finally, the residues were coloured according to their threading energy, which makes it a lot easier to read. The results from this study are shown in tables 7-8. It is obvious that with KAS I/II most of the residues have a threading energy, which is positive, i.e. above zero. However, there are only four points where the residues are really "unhappy" from their surroundings. Those residues are ASN118, VAL349, PHE350 and ASP404. Special attention needs to be paid to those residues, as they are considered to be "high risk" residues to cause problems to the model later on. Same way, the threading energy of the KAS III enzyme showed that there another four residues, which are really unhappy from their surroundings. Those are ILE80, ASN313, ILE314 and PRO315.

Table 7. Curve showing the threading energy of the KAS I/II enzyme (aligned with its template).

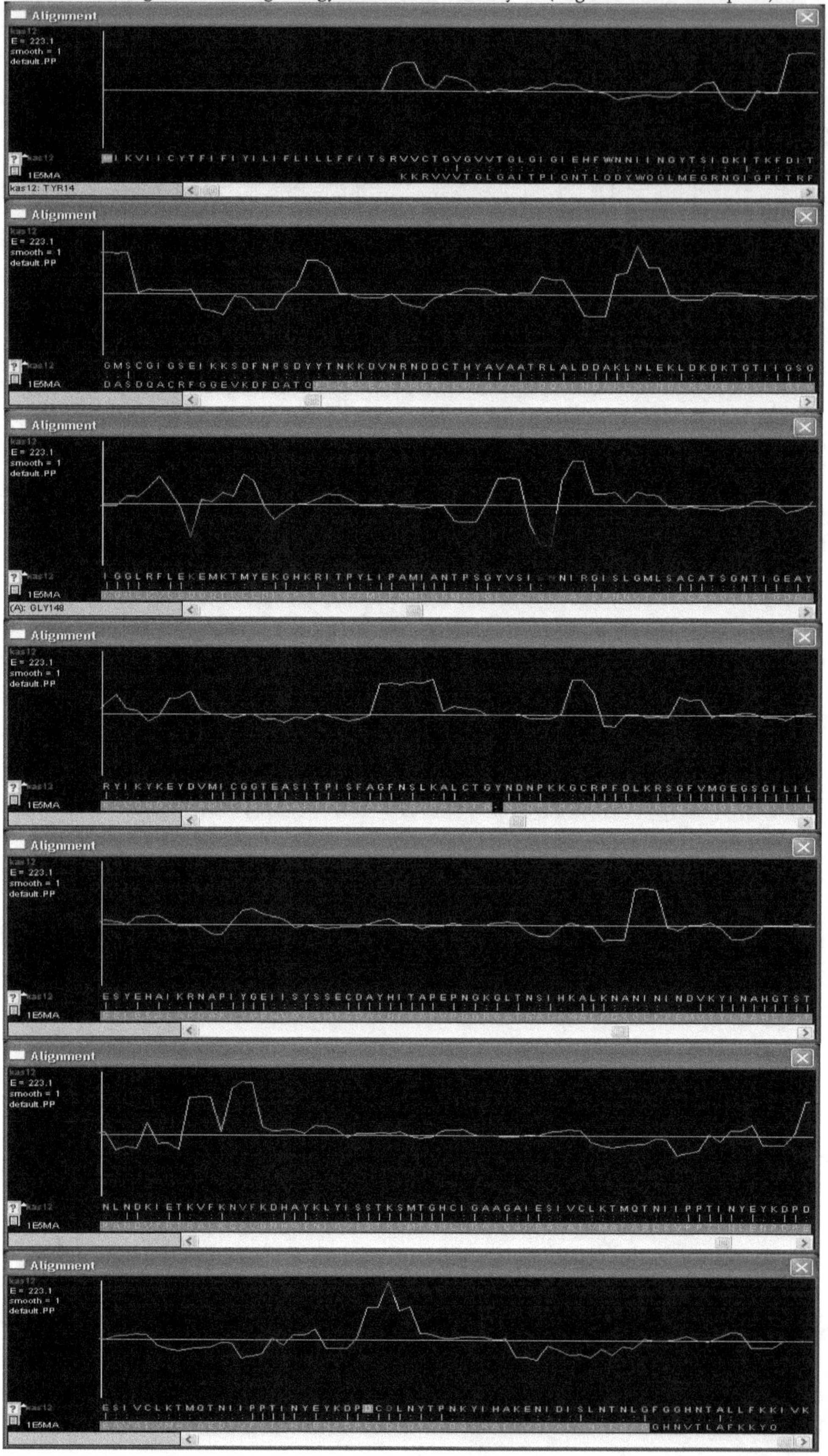

Table 8. Curve showing the threading energy of the KAS III enzyme (aligned with its template).

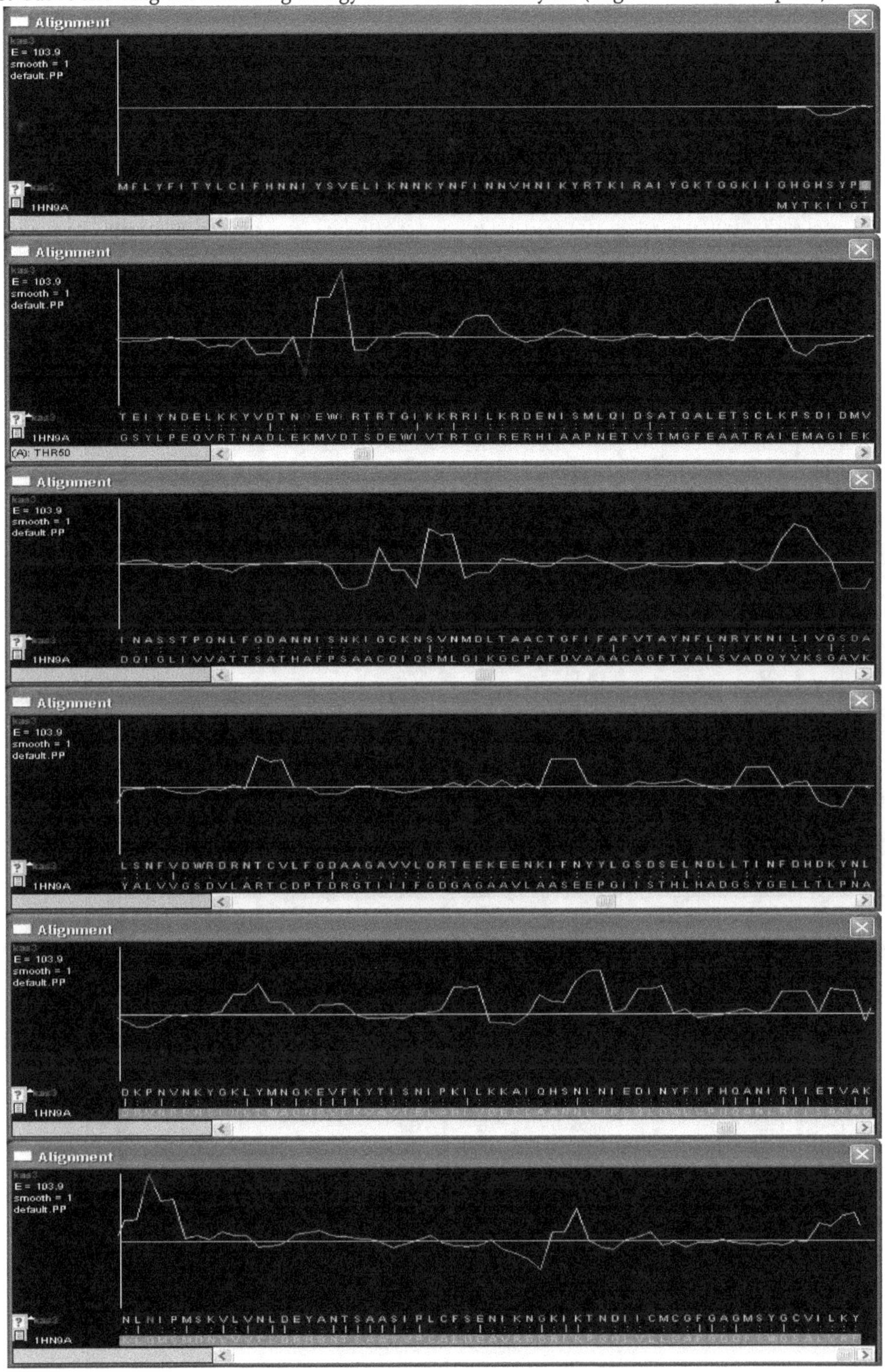

Table 9. The threading energy of the KAS I/II enzyme (coloured by residue).

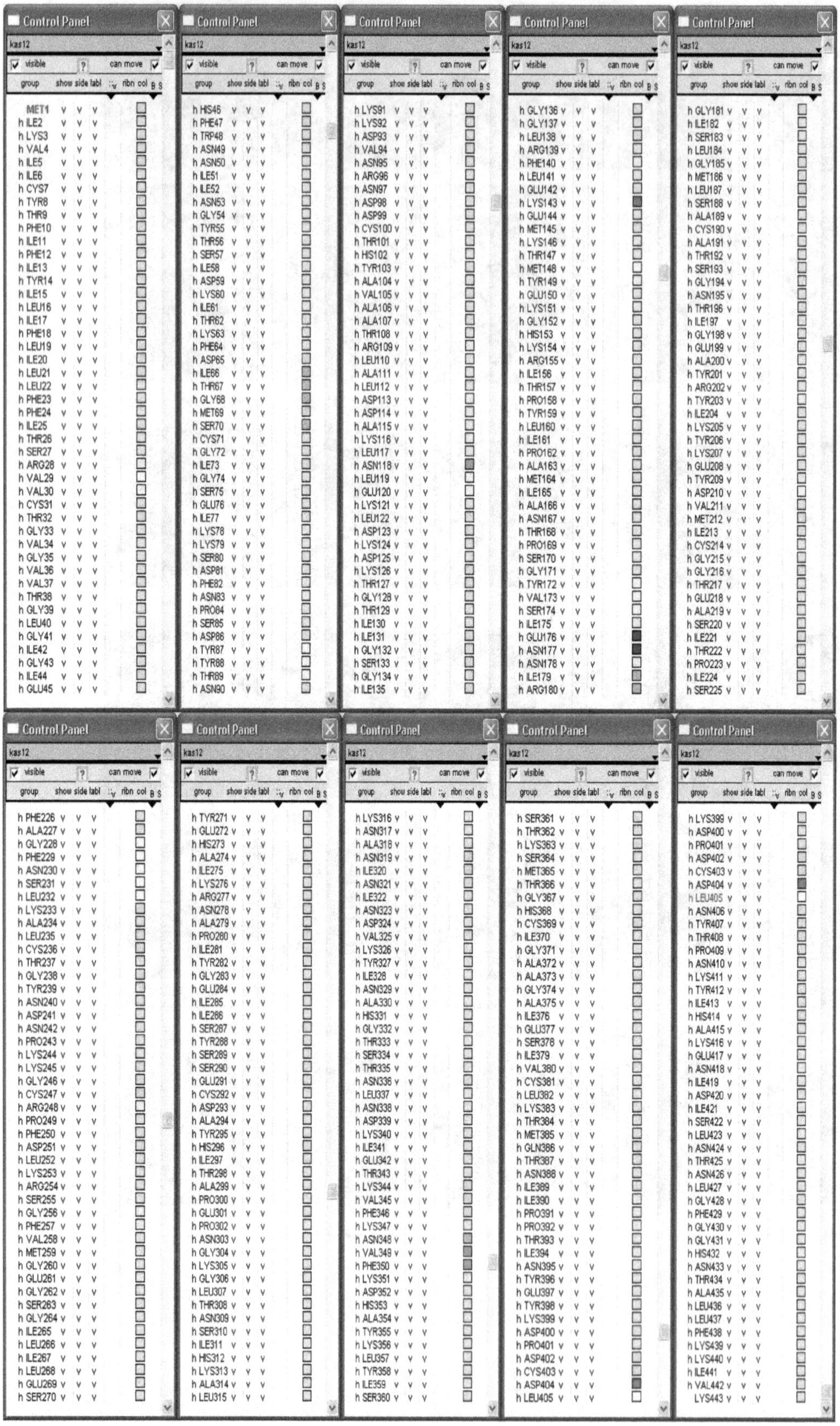

Table 10. The threading energy of the KAS III enzyme (coloured by residue).

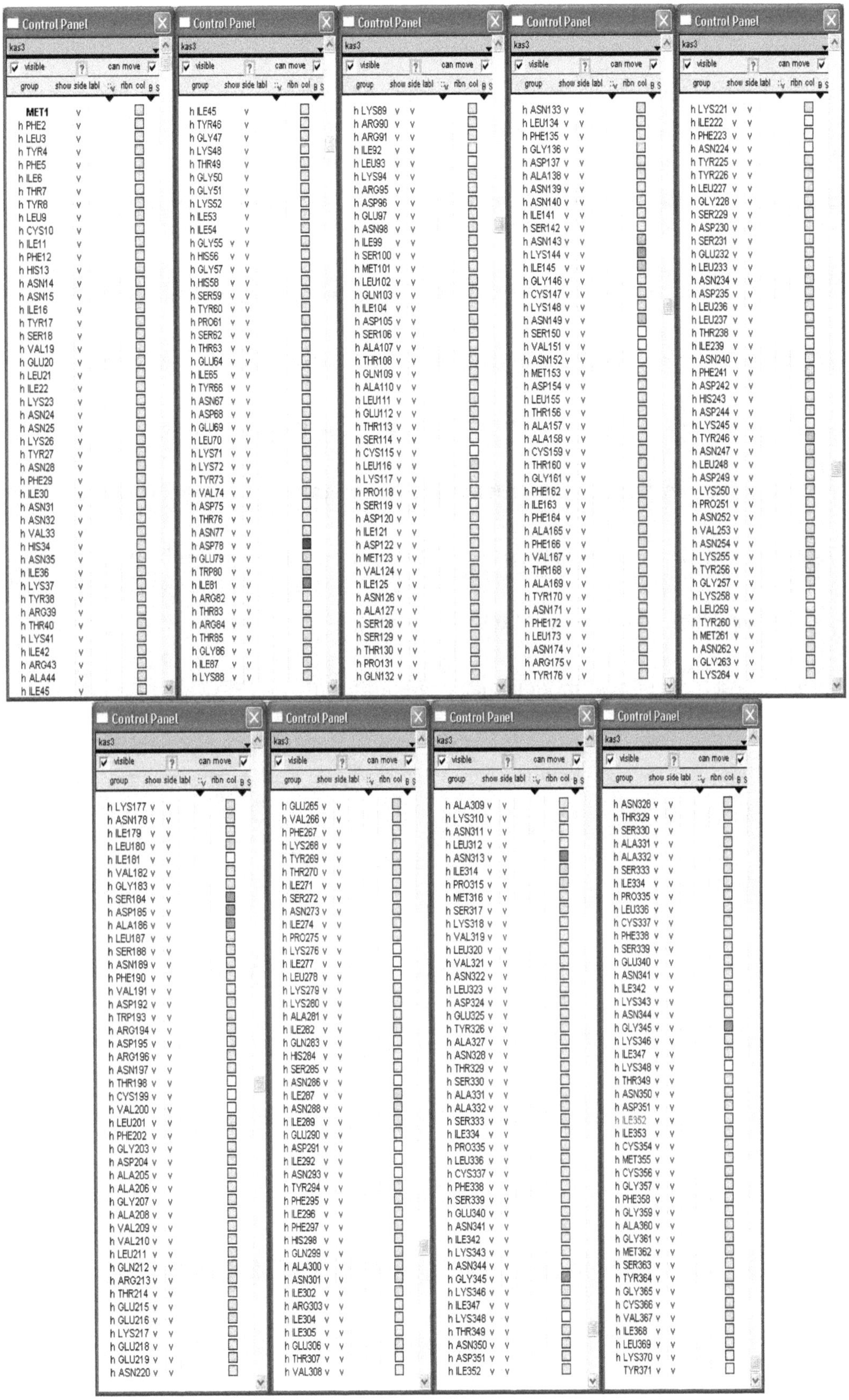

2.4.4 Homology Modelling

The residues mentioned above, which appear in red in the figures above, are the high-possibility residues to cause problem when building the model. The next step is to go back to Composer, import the sequence and try to build a model. The first problem that had to be overcome was the loops that were missing from the model. Unfortunately, the 1800 structure database was not enough to find all the loops that were missing. An attempt was made to import some new structure into the Composer's database, but this did not help much either. Fortunately, another computer running Composer was found whose database was several thousand structures. That solved the loop problem and a model was made at last. That was for both the KAS I/II and KAS III sequences. Below are two sets of figures for each of the two models that were built with COMPOSER.

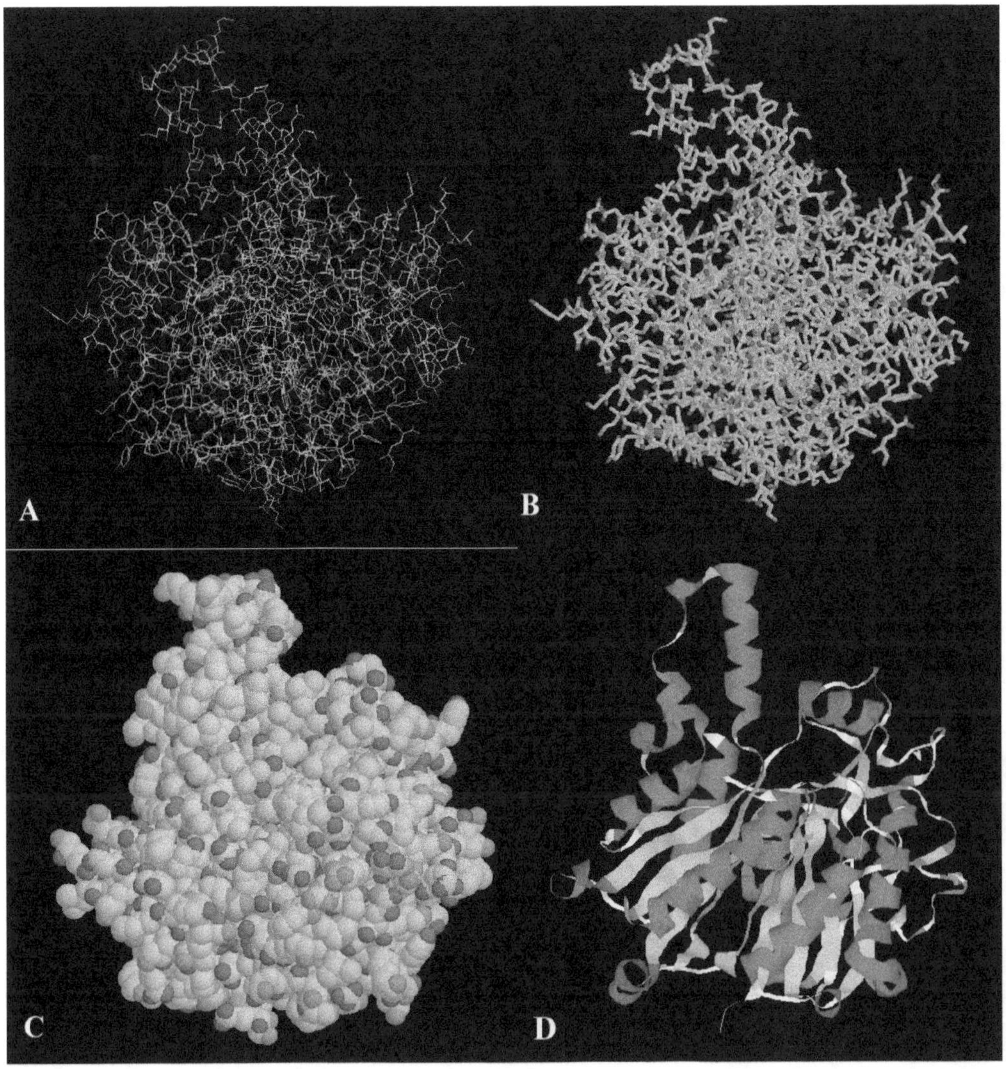

Figure 25. The Structure of the KAS I/II model, built by COMPOSER by Homology Modelling. A: the model in wireframe display, B: the model in stick display, C: the model in spacefill display and D: the model in cartoon display, where a-helices are in pink and β-sheet.

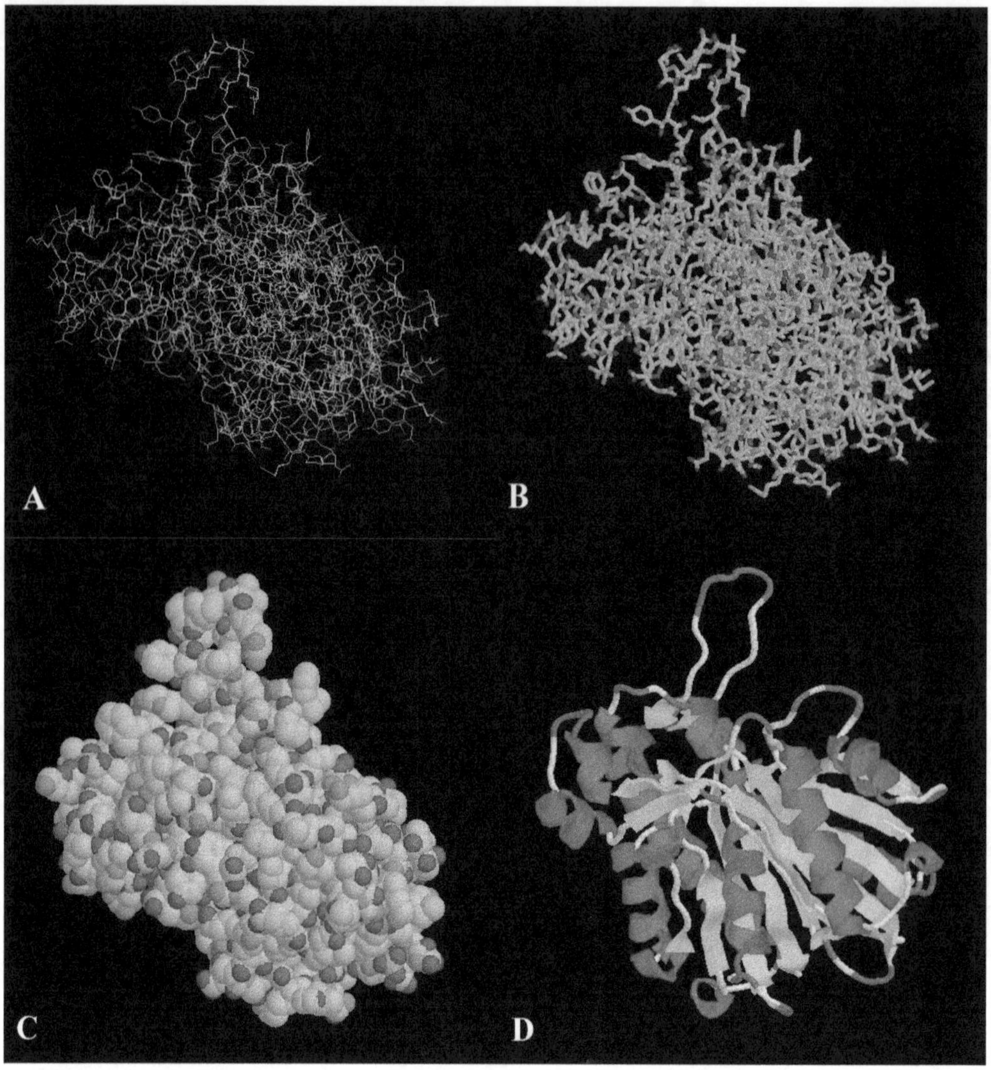

Figure 26. The Structure of the KAS I/II model, built by COMPOSER by Homology Modelling. A: the model in wireframe display, B: the model in stick display, C: the model in spacefill display and D: the model in cartoon display, where a-helices are in pink and β-sheet.

There are two ways to build a 3D model by homology modelling. The first way is based on strict superimposition. The second method is based in geometric restraints, which includes analogy to the molecular replacement and distance geometry methodologies. Composer takes into account both of them. There are a number of algorithms that have been developed in order to help make a model. The most popular ones are the Composer, Jig-Saw and SwissProt. In this experiment both the Composer and SwissProt were used with the latter being a bit more efficient at building the SCRs. Usually the construction of SCRs uses similar SCRs in sequence from proteins that are homologous to each other. That is done in order to define a new core out of a multiple alignment. So, first a framework from SCRs is being made by superimposing known structures. Then the chosen homologue sequence is superimposed on the unknown sequence in the main chain of the framework, which has been averaged. Then the anchored stumps on the framework are found and compatible structures are placed on them. Finally, the side chains are modelled. Now, the quality of the models had to be checked and confirmed. There is a variety of software that checks the quality of the model on behalf of the scientist.

2.4.5 Model Evaluation

PROCHECK is the software package that was used on both KAS I/II and KAS III models. The models were checked in terms of their stereochemistry. PROCHECK found no problems in both models. The results that PROCHECK returned are fully acceptable for both KAS I/II and KAS III models. The table below summarises the result that PROCHECK prodused for both the models and their templates.

Table 11. Summary of the results that PROCHECK produced for the KAS I/II and KAS III models and their templates.

		Model 1		Model 2	
		KAS I/II	**1E5M**	**KAS III**	**1HN9**
Residues		411	411	317	317
Ramachandran	Core	88.2	88.2	92.4	92.4
	Allowed	11.3	11.3	6.9	6.9
	General	0.3	0.3	0.7	0.7
	disallowed	0.3	0.3	0.0	0.0
Bond Lengths		100	100	100	100
Bond Angles		90.6	90.6	89.7	89.7
Planar Groups		97.3	97.3	89.5	89.5

The most important numbers in the PROCHECK calculation are the percentage of bond length and bond angle that is within the acceptable limit. In the case of KAS I/II the percentages of bond length and bond angle within the desired range were 100% and 90.6% respectively. As far as the KAS III model is concerned the percentages of bond length and bond angle within the desired range were 100% and 89.7% respectively. All those percentages are considered to be satisfactory and acceptable. The tables below show the results of PROCHECK on both the KAS I/II and KAS III Plasmodium models and their templates. Figures 28 and 29 show the ramachandran plots obtained for the KAS I/II and KAS III Plasmodium models.

Figure 27. Secondary structure and Ramachandran regions from PROCHECK for the KAS I/II Plasmodium model.

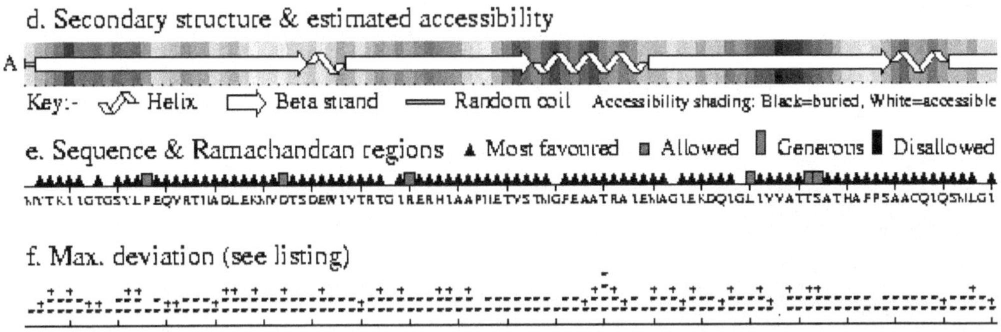

Figure 28. The ramachandran plot that PROCHECK produced for the KAS I/II Plasmodium model.

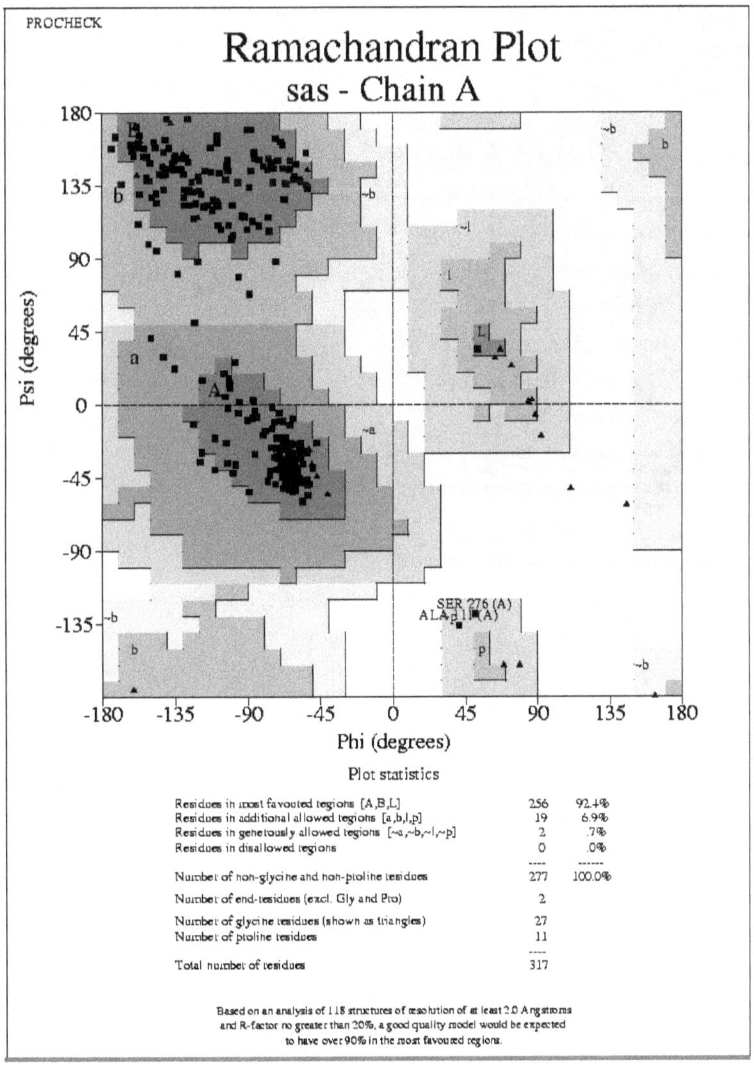

Figure 29. The ramachandran plot that PROCHECK produced for the KAS III Plasmodium model.

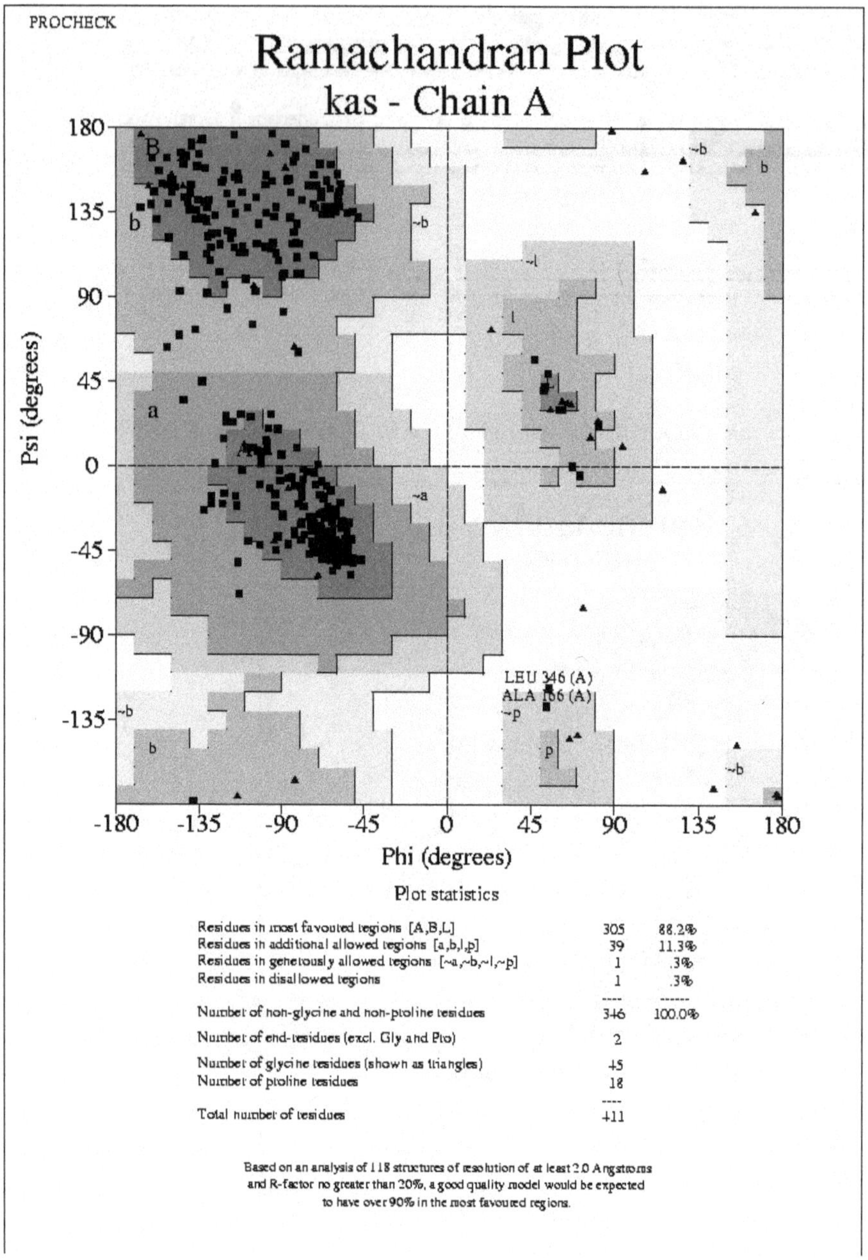

2.5 Conclusions:

The protein of KAS consists of four identical subunits with an active site on each subunit. It was found that the way the ligand binds is different among the four active sites of the protein. Thiolactomycin binds by establishing different sets of interactions among the active sites of the protein. The residues of the active site though are the same in all four active sites of the protein. On the other hand Cerulenin not only binds by establishing different interactions among the active sites, but it also uses one more residue in subunit B. As a result, it is not only the type of interactions that varies from active site to active site, but it is also the number of residues that varies too. Moreover superimpositions of the empty active site over the same active site with the ligand present, reveal significant active-site residue movement when the ligand binds. That stands the same for both Thiolactomycin and Cerulenin. It was found that the distance that the active site's residues move when the ligand binds, depends on the exposed contact surface of each residue. Residues with large contact surface area moved a lot, whereas residues with little exposed contact surface area either moved little or not at all.

The sequence alignment results were very good, since both identity percentages (for KAS I/II and KAS III) were above the limit of 30%, which actually got up to 45%. The key residues of the sequence were doubled-checked by eye and it was confirmed that the alignment had been done correctly. Key residues are considered to be those that make the active site of the protein (especially the two histidines, which establish H-bonds) and those that are generally conserved (evolutionary). Building the model was a quite straightforward procedure and the quality of the models was assessed with PROCHECK. The results of the quality check of the models proved that the models are very accurate and reliable. The quality of the models was actually as good as it could be, given the quality of the templates that were chosen.

In the homology modelling of both enzymes it was decided that the larger the COMPOSER's database the better the model, since the missing loops are retrieved out of the program's pdb database. Finally the models are considered to be of high standards and suitable for future docking and drug-design experiments.

2.6 Future Work:

An area that could be further investigated is the structural differences of the active sites of the four subunits on the same protein. For example, the k_m of each active site could be measured and that would provide vital information on whether there is a matter of cooperativity or not. Then further studies could be made with only one ligand docked, then with another one and so on.

The ultimate aim of such an experiment is to create the model of the proteins structure out of its sequence and then to design the inhibitor that will bind to the model's active site too. This latter part has not been done yet. A first approach would involve the docking of Thiolactomycin and Cerulenin into the two KAS models. Then, with the aid of ligand-designing programs, such as Chemera or Ligand-X a variety of compounds could be tested in terms of binding efficiency to the active site of the model. The best compounds could then be further improved in terms of energy minimization and structural conformation. This route is a lot easier from starting from scratch. Still, if this method does not work, then the ligand will have to be designed from the data of the active site studies. This would include the structural details of the active site, the residues that it is made of and finally the type of interactions that take place when the ligand binds. Finally, the docking of the designed compound using docking software (e.g. Dock, AutoDock, Gold) would produce the inhibitor of that particular protein.

2.7 Appendix.

LPC ANALYSIS.

Contacts of ligand CER 413A in PDB entry 1B3N

Table II
Residues in contact with ligand CER 413A in PDB entry 1B3N
Legend:
Dist - nearest distance (Å) between atoms of the ligand and the residue
Surf - contact surface area (Å^2) between the ligand and the residue
HB - hydrophilic-hydrophilic contact (hydrogen bond)
Arom - aromatic-aromatic contact
Phob - hydrophobic-hydrophobic contact
DC - hydrophobic-hydrophilic contact (destabilizing contact)
+/- - indicates presence/absence of a specific contacts
* - indicates residues contacting ligand by their side chain (including CA atoms)

		Specific contacts			
Residue	Dist	Surf	HB	Arom	Phob	DC

Residue	Dist	Surf	HB	Arom	Phob	DC
107A GLY*	3.7	9.2	-	-	-	+
108A ILE*	3.3	55.2	-	-	+	+
111A LEU*	3.7	11.7	-	-	+	-
162A ALA*	3.2	35.0	-	-	+	+
163A CYS*	1.8	53.9	-	-	-	+
191A GLU*	3.6	11.2	-	-	-	+
193A ALA*	3.6	16.8	-	-	+	-
197A LEU	4.5	3.1	-	-	-	+
198A GLY*	3.4	29.8	-	-	-	+
201A GLY*	5.8	1.1	-	-	-	-
202A PHE*	3.4	47.3	-	-	+	+
229A PHE*	5.4	8.0	-	-	+	-
270A THR	5.1	2.2	+	-	-	-
303A HIS*	2.9	29.5	+	-	-	-
305A THR*	4.4	1.3	-	-	-	+
340A HIS*	3.0	10.0	+	-	+	+
342A LEU*	3.7	23.6	-	-	+	-
398A PHE	3.4	3.6	-	-	-	+
399A GLY*	3.3	3.9	-	-	-	-
400A PHE*	2.6	85.5	+	-	+	-

Table III
List of putative hydrogen bonds between ligand CER 413A and protein in PDB entry 1B3N
Legend:
N - ligand atom number in PDB entry
Dist - distance (Å) between the ligand and protein atoms
Surf - contact surface area (Å^2) between the ligand and protein atoms

Ligand atom			Protein atom				
N	Name	Class	Residue	Name	Class	Dist	Surf
2	N1	I	PHE 400A	N	III	4.3	2.2

2	N1	I	HIS 303A	NE2	I	4.8	0.2
2	N1	I	THR 270A	O	II	5.1	2.2
5	O2	II	HIS 303A	NE2	I	2.9	29.3
5	O2	II	HIS 340A	NE2	I	3.0	4.7
7	O3	I	PHE 400A	N	III	2.6	29.3

Table IV
Full list of atomic contacts with ligand CER 413A in PDB entry 1B3N
Total number of contacts is 95
Legend:
N - ligand atom number in PDB entry
Dist - distance (A) between the ligand and protein atoms
Surf - contact surface area (A**2) between the ligand and protein atoms
* - indicates destabilizing contacts

Ligand atom			Protein atom				
N	Name	Class	Residue	Name	Class	Dist	Surf
1	C1	VI	CYS 163A	SG	VI	2.8	0.4
1	C1	VI	CYS 163A	CB	IV	3.5	1.6
1	C1	VI	HIS 340A	NE2	I	3.8	2.7
1	C1	VI	PHE 398A	O	II	4.0	2.9
1	C1	VI	HIS 340A	CD2	V	4.5	2.0
1	C1	VI	THR 305A	CG2	IV	5.4	0.4
1	C1	VI	PHE 229A	CE2	V	5.5	2.2
1	C1	VI	PHE 229A	CD2	V	5.8	0.4
2	N1	I	PHE 400A	CD1	V	3.4	15.2
2	N1	I	PHE 400A	CE1	V	3.4	6.1
2	N1	I	PHE 400A	N	III	4.3	2.2
2	N1	I	GLY 399A	CA	VI	4.5	1.4
2	N1	I	GLY 399A	C	VI	4.7	0.4
2	N1	I	HIS 303A	NE2	I	4.8	0.2
2	N1	I	THR 270A	O	II	5.1	2.2
2	N1	I	PHE 398A	C	VI	5.3	0.2
2	N1	I	PHE 229A	CE2	V	5.4	3.8
2	N1	I	PHE 229A	CD2	V	5.8	0.4
3	O1	II	PHE 400A	CD2	V	3.2	16.6
3	O1	II	PHE 400A	CE2	V	3.3	4.9
3	O1	II	PHE 202A	CZ	V	3.4	8.5
4	C2	IV	CYS 163A	SG	VI	1.8	35.0
4	C2	IV	HIS 340A	NE2	I	4.1	0.2*
4	C2	IV	LEU 342A	CD2	IV	4.3	4.3
4	C2	IV	HIS 340A	CD2	V	4.7	0.2
4	C2	IV	PHE 202A	CZ	V	5.4	0.4
4	C2	IV	PHE 202A	CE2	V	5.4	0.9
4	C2	IV	PHE 229A	CE2	V	5.6	1.1
5	O2	II	HIS 303A	NE2	I	2.9	29.3
5	O2	II	CYS 163A	SG	VI	2.9	6.2
5	O2	II	HIS 340A	NE2	I	3.0	4.7
5	O2	II	CYS 163A	CB	IV	3.2	1.0*
5	O2	II	PHE 398A	O	II	3.4	0.5*
5	O2	II	HIS 340A	CD2	V	3.9	0.2
5	O2	II	THR 305A	CG2	IV	4.4	0.9*
6	C3	VI	PHE 400A	CD1	V	3.4	6.7

6	C3	VI	PHE	400A	CE1	V	3.8	0.7
6	C3	VI	PHE	400A	CZ	V	4.3	0.2
6	C3	VI	PHE	202A	CE2	V	5.0	0.2
7	O3	I	PHE	400A	N	III	2.6	29.3
7	O3	I	ALA	162A	CB	IV	3.2	6.9*
7	O3	I	GLY	399A	CA	VI	3.3	2.1
8	C4	VIII	ALA	162A	CB	IV	3.8	1.1
8	C4	VIII	PHE	400A	CG	V	3.9	1.6
8	C4	VIII	PHE	202A	CZ	V	4.1	3.4
9	C5	IV	ALA	162A	CB	IV	3.3	10.3
9	C5	IV	CYS	163A	N	III	3.4	6.1*
9	C5	IV	CYS	163A	SG	VI	3.4	3.6
9	C5	IV	ALA	162A	CA	VII	3.5	0.9
9	C5	IV	ALA	162A	C	VI	3.7	0.9
9	C5	IV	LEU	342A	CD2	IV	4.4	2.0
9	C5	IV	LEU	342A	CB	IV	4.7	0.4
10	C6	IV	ALA	162A	CB	IV	3.6	11.9
10	C6	IV	ALA	162A	CA	VII	3.7	2.9
10	C6	IV	GLY	107A	CA	VI	4.8	0.2
10	C6	IV	GLY	107A	C	VI	5.0	0.7
10	C6	IV	GLY	107A	O	II	5.0	0.7*
10	C6	IV	PHE	400A	CD2	V	5.1	1.8
10	C6	IV	PHE	400A	CB	IV	5.5	0.2
11	C7	IV	PHE	202A	CZ	V	3.9	6.5
11	C7	IV	LEU	342A	CD1	IV	4.5	1.8
12	C8	IV	ILE	108A	N	III	3.7	17.7*
12	C8	IV	GLY	107A	C	VI	3.7	4.0
12	C8	IV	ILE	108A	CB	IV	3.8	1.1
12	C8	IV	GLY	107A	CA	VI	3.8	3.6
12	C8	IV	GLU	191A	OE2	II	4.2	2.0*
12	C8	IV	LEU	342A	CD1	IV	4.4	0.4
12	C8	IV	LEU	342A	CB	IV	4.6	0.4
12	C8	IV	ILE	108A	CD1	IV	5.1	1.1
13	C9	IV	ILE	108A	CG2	IV	3.4	12.3
13	C9	IV	GLU	191A	OE1	II	3.6	9.0*
13	C9	IV	LEU	342A	CD1	IV	3.7	14.1
13	C9	IV	GLU	191A	CD	VI	3.9	0.2
13	C9	IV	ALA	193A	CB	IV	4.1	0.9
14	C10	IV	ILE	108A	CG2	IV	3.3	13.7
14	C10	IV	ILE	108A	CB	IV	3.4	3.1
14	C10	IV	ILE	108A	CG1	IV	4.2	0.7
14	C10	IV	ILE	108A	CD1	IV	4.4	2.0
15	C11	IV	PHE	202A	CE1	V	3.4	17.0
15	C11	IV	ALA	193A	CB	IV	3.6	15.9
15	C11	IV	ILE	108A	CG2	IV	4.1	0.2
15	C11	IV	GLY	198A	CA	VI	4.3	0.2
16	C12	IV	GLY	198A	CA	VI	3.4	28.9
16	C12	IV	LEU	111A	CD1	IV	3.7	11.7
16	C12	IV	PHE	202A	CE1	V	3.9	8.1
16	C12	IV	GLY	198A	O	II	4.2	0.7*
16	C12	IV	PHE	202A	CD1	V	4.3	0.4
16	C12	IV	LEU	197A	O	II	4.5	3.1*
16	C12	IV	PHE	202A	CZ	V	4.6	0.7
16	C12	IV	ILE	108A	CB	IV	5.1	0.2
16	C12	IV	ILE	108A	CG1	IV	5.2	1.6
16	C12	IV	ILE	108A	CD1	IV	5.2	1.3
16	C12	IV	PHE	202A	N	III	5.7	1.1*
16	C12	IV	GLY	201A	CA	VI	5.8	0.9

16 C12 IV GLY 201A C VI 6.1 0.2

Table VI
Normalised complementarity as a function of atomic substitution for ligand CER 413A in PDB entry 1B3N

Legend:

N - ligand atom number in PDB entry

Bold - indicates atomic substitution which could stabilize the complex

Italics - indicates atomic substitution which could destabilize the complex

Ligand atom			Atom class							
N	Type	Class	I	II	III	IV	V	VI	VII	VIII
1	C1	VI	0.71	0.69	0.71	0.69	0.71	0.71	0.71	0.70
2	N1	I	0.71	0.70	0.70	0.69	0.71	0.71	0.70	0.70
3	O1	II	0.71	0.71	0.71	0.71	0.71	0.71	0.71	0.71
4	C2	IV	0.70	0.70	0.70	0.71	0.71	0.71	0.71	0.71
5	O2	II	0.72	0.71	0.72	0.58	0.72	0.72	0.72	0.72
6	C3	VI	0.71	0.71	0.71	0.71	0.71	0.71	0.71	0.71
7	O3	I	0.71	0.71	0.59	0.62	0.74	0.74	0.62	0.74
8	C4	VIII	0.71	0.71	0.71	0.71	0.71	0.71	0.71	0.71
9	C5	IV	0.69	0.69	0.66	0.71	0.74	0.74	0.71	0.74
10	C6	IV	0.67	0.66	0.65	0.71	0.72	0.72	0.70	0.71
11	C7	IV	0.71	0.71	0.71	0.71	0.71	0.71	0.71	0.71
12	C8	IV	0.78	0.77	0.71	0.71	0.80	0.80	0.72	0.79
13	C9	IV	0.64	0.60	0.64	0.71	0.75	0.75	0.75	0.71
14	C10	IV	0.63	0.63	0.63	0.71	0.71	0.71	0.71	0.71
15	C11	IV	0.65	0.65	0.65	0.71	0.71	0.71	0.71	0.71
16	C12	IV	0.67	0.66	0.67	0.71	0.73	0.73	0.73	0.72

Legitimacy depends on the hydrophobic-hydrophilic properties of the contacting atoms. In order to define it, for each inter-atomic contact, eight atom classes have been introduced:

I Hydrophilic - N and O that can donate and accept hydrogen bonds (e.g., oxygen of hydroxyl group of Ser. or Thr)
II Acceptor - N or O that can only accept a hydrogen bond
III Donor - N that can only donate a hydrogen bond
IV Hydrophobic - Cl, Br, I and all C atoms that are not in aromatic rings and do not have a covalent bond to a N or O atom
V Aromatic - C in aromatic rings irrespective of any other bonds formed by the atom
VI Neutral - C atoms that have a covalent bond to at least one atom of class I or two or more atoms from class II or III; atoms; S, F, P, and metal atoms in all cases
VII Neutral-donor - C atoms that have a covalent bond with only one atom of class III

VIII Neutral-acceptor - C atoms that have a covalent bond with only one atom of class II

For each pair of contacts the state of legitimacy is shown below:

Legend:
+, legitimate
-, illegitimate

Atomic class	I	II	III	IV	V	VI	VII	VIII
I (Hydrophilic)	+	+	+	-	+	+	+	+
II (Acceptor)	+	-	+	-	+	+	+	-
III (Donor)	+	+	-	-	+	+	-	+
IV (Hydrophobic)	-	-	-	+	+	+	+	+
V (Aromatic)	+	+	+	+	+	+	+	+
VI (Neutral)	+	+	+	+	+	+	+	+
VII (Neutral-donor)	+	+	-	+	+	+	-	+
VIII (Neutral-acceptor)	+	-	+	+	+	+	+	-

Contacts of ligand DAO 901A in PDB entry 1EK4

Table II
Residues in contact with ligand DAO 901A in PDB entry 1EK4
Legend:
Dist - nearest distance (Å) between atoms of the ligand and the residue
Surf - contact surface area (Å^2) between the ligand and the residue
HB - hydrophilic-hydrophilic contact (hydrogen bond)
Arom - aromatic-aromatic contact
Phob - hydrophobic-hydrophobic contact
DC - hydrophobic-hydrophilic contact (destabilizing contact)
+/- - indicates presence/absence of a specific contacts
* - indicates residues contacting ligand by their side chain
(including CA atoms)

Residue	Dist	Surf	Specific contacts: HB	Arom	Phob	DC
106A GLY*	3.5	37.9	-	-	-	+
107A GLY*	3.9	11.9	-	-	-	+
110A PRO*	4.2	8.5	-	-	+	-
162A ALA*	3.3	21.3	-	-	+	+
163A SER*	1.3	62.9	+	-	-	+
191A GLU*	4.7	0.7	-	-	-	+
197A MET*	3.4	43.1	-	-	+	+
200A GLU*	3.8	37.9	-	-	+	+
201A PHE*	3.6	36.1	-	-	+	-
229A PHE*	5.1	2.7	-	-	+	-
298A HIS*	5.4	1.3	-	-	-	-
333A HIS*	3.9	12.1	-	-	+	+
335A LEU*	4.1	26.2	-	-	+	-
390A PHE	4.4	0.2	-	-	-	+
391A GLY*	3.1	1.8	-	-	-	-
392A PHE*	3.0	56.6	+	-	+	-
113B GLN*	4.0	10.8	-	-	-	+
134B VAL*	3.8	23.3	-	-	+	+
137B ALA*	3.8	14.8	-	-	+	+
138B MET*	3.6	44.6	-	-	+	+

Table III
List of putative hydrogen bonds between ligand DAO 901A and protein in PDB entry 1EK4
Legend:
N - ligand atom number in PDB entry
Dist - distance (Å) between the ligand and protein atoms
Surf - contact surface area (Å^2) between the ligand and protein atoms

Ligand atom: N	Name	Class	Protein atom: Residue	Name	Class	Dist	Surf

1	O1	II	SER 163A	N	III	2.9	7.1
1	O1	II	PHE 392A	N	III	3.0	23.4

Table IV
Full list of atomic contacts with ligand DAO 901A in PDB entry 1EK4
Total number of contacts is 84
Legend:
N - ligand atom number in PDB entry
Dist - distance (A) between the ligand and protein atoms
Surf - contact surface area (A**2) between the ligand and protein atoms
* - indicates destabilizing contacts

Ligand atom			Protein atom				
N	Name	Class	Residue	Name	Class	Dist	Surf
1	O1	II	SER 163A	CB	VI	2.8	6.9
1	O1	II	SER 163A	N	III	2.9	7.1
1	O1	II	PHE 392A	N	III	3.0	23.4
1	O1	II	GLY 391A	CA	VI	3.1	1.6
1	O1	II	ALA 162A	CB	IV	3.3	0.2*
1	O1	II	PHE 392A	CD2	V	3.7	0.9
2	C1	VIII	SER 163A	OG	I	1.3	35.0
2	C1	VIII	SER 163A	CB	VI	2.4	6.3
2	C1	VIII	SER 163A	N	III	3.0	3.4
2	C1	VIII	HIS 333A	NE2	I	3.9	1.8
2	C1	VIII	GLY 391A	CA	VI	4.1	0.2
2	C1	VIII	PHE 392A	CD2	V	4.3	0.9
2	C1	VIII	PHE 390A	O	II	4.4	0.2*
2	C1	VIII	PHE 392A	CE2	V	4.5	0.2
2	C1	VIII	HIS 298A	CE1	V	5.4	0.7
2	C1	VIII	HIS 298A	NE2	I	5.7	0.7
3	C2	IV	SER 163A	OG	I	2.5	1.8*
3	C2	IV	PHE 392A	CE2	V	3.9	13.9
3	C2	IV	PHE 392A	CZ	V	4.0	2.0
3	C2	IV	HIS 333A	NE2	I	4.1	8.7*
3	C2	IV	LEU 335A	CD2	IV	4.1	5.8
3	C2	IV	PHE 201A	CE1	V	4.2	2.7
3	C2	IV	HIS 333A	CD2	V	4.8	1.6
3	C2	IV	PHE 229A	CZ	V	5.1	2.0
3	C2	IV	PHE 229A	CE2	V	5.2	0.7
4	C3	IV	PHE 201A	CE1	V	3.6	15.0
4	C3	IV	PHE 392A	CD1	V	3.7	14.8
4	C3	IV	PHE 392A	CG	V	3.9	0.4
4	C3	IV	ALA 162A	CB	IV	4.2	0.2
5	C4	IV	PHE 201A	CZ	V	3.7	5.6
5	C4	IV	SER 163A	OG	I	3.7	0.7*
5	C4	IV	ALA 162A	CB	IV	4.0	1.6
5	C4	IV	SER 163A	N	III	4.1	1.8*
5	C4	IV	LEU 335A	CD2	IV	4.1	7.2
5	C4	IV	ALA 162A	C	VI	4.2	0.4
5	C4	IV	LEU 335A	CB	IV	4.5	0.7
6	C5	IV	ALA 162A	CB	IV	3.8	11.2

6	C5	IV	PHE	201A	CZ	V	4.0	1.3
6	C5	IV	MET	138B	SD	VIII	4.1	11.9
6	C5	IV	VAL	134B	CG1	IV	4.4	1.6
7	C6	IV	GLY	106A	O	II	3.5	14.1*
7	C6	IV	GLY	106A	C	VI	3.6	4.0
7	C6	IV	GLY	106A	CA	VI	3.7	5.8
7	C6	IV	ALA	162A	CA	VII	3.9	7.0
7	C6	IV	MET	138B	CB	IV	4.1	1.1
7	C6	IV	ALA	162A	C	VI	4.4	0.7
7	C6	IV	LEU	335A	CB	IV	4.8	1.1
8	C7	IV	GLY	106A	C	VI	3.7	9.0
8	C7	IV	GLY	107A	N	III	3.9	1.3*
8	C7	IV	PHE	201A	CZ	V	4.0	4.9
8	C7	IV	GLY	106A	CA	VI	4.1	1.1
8	C7	IV	LEU	335A	CD1	IV	4.3	11.4
8	C7	IV	MET	197A	SD	VIII	4.5	0.7
8	C7	IV	GLU	191A	OE2	II	4.7	0.7*
9	C8	IV	MET	138B	CB	IV	3.6	10.3
9	C8	IV	GLY	106A	O	II	3.8	3.8*
9	C8	IV	PHE	201A	CZ	V	4.2	2.7
9	C8	IV	VAL	134B	CG1	IV	4.5	0.9
10	C9	IV	MET	197A	CG	IV	3.4	19.5
10	C9	IV	MET	197A	CE	IV	3.5	5.4
10	C9	IV	GLY	107A	CA	VI	4.0	0.2
10	C9	IV	PHE	201A	CE2	V	4.1	2.9
11	C10	IV	MET	197A	CG	IV	3.9	6.5
11	C10	IV	MET	197A	CE	IV	3.9	4.9
11	C10	IV	GLY	107A	CA	VI	4.0	9.9
11	C10	IV	PRO	110A	CG	IV	4.2	8.5
11	C10	IV	GLN	113B	OE1	II	4.2	1.1*
11	C10	IV	MET	138B	CA	VII	4.3	2.5
11	C10	IV	ALA	137B	O	II	4.5	2.2*
11	C10	IV	GLY	107A	O	II	4.7	0.4*
12	C11	IV	MET	138B	N	III	3.8	18.8*
12	C11	IV	ALA	137B	CB	IV	3.8	10.3
12	C11	IV	VAL	134B	O	II	4.0	4.5*
12	C11	IV	GLN	113B	OE1	II	4.1	0.7*
13	C12	IV	GLU	200A	CB	IV	3.8	29.4
13	C12	IV	VAL	134B	CA	VII	3.8	12.8
13	C12	IV	GLU	200A	CG	IV	4.0	5.4
13	C12	IV	GLN	113B	OE1	II	4.0	9.0*
13	C12	IV	VAL	134B	CG2	IV	4.0	3.6
13	C12	IV	GLU	200A	OE1	II	4.1	3.1*
13	C12	IV	MET	197A	O	II	4.3	5.2*
13	C12	IV	MET	197A	CG	IV	4.3	0.9
13	C12	IV	ALA	137B	CB	IV	4.4	2.2
13	C12	IV	PHE	201A	CE2	V	4.7	0.9

Table V

Table VI
Normalised complementarity as a function of atomic substitution for ligand DAO 901A in PDB entry 1EK4

Legend:

N - ligand atom number in PDB entry

Bold - indicates atomic substitution which could stabilize the complex

Italics - indicates atomic substitution which could destabilize the complex

Ligand atom			Atom class							
N	Type	Class	I	II	III	IV	V	VI	VII	VIII
1	O1	II	0.67	0.67	0.53	0.53	0.67	0.67	0.53	0.67
2	C1	VIII	0.67	0.67	0.65	0.48	0.67	0.67	0.65	0.67
3	C2	IV	0.69	0.69	0.69	0.67	0.71	0.71	0.71	0.71
4	C3	IV	0.66	0.66	0.66	0.67	0.67	0.67	0.67	0.67
5	C4	IV	0.63	0.63	0.63	0.67	0.68	0.68	0.67	0.68
6	C5	IV	0.61	0.56	0.61	0.67	0.67	0.67	0.67	0.61
7	C6	IV	0.72	0.66	0.69	0.67	0.73	0.73	0.70	0.67
8	C7	IV	0.62	0.62	0.62	0.67	0.68	0.68	0.67	0.67
9	C8	IV	0.63	0.62	0.63	0.67	0.68	0.68	0.68	0.67
10	C9	IV	0.55	0.55	0.55	0.67	0.67	0.67	0.67	0.67
11	C10	IV	0.59	0.58	0.58	0.67	0.68	0.68	0.67	0.67
12	C11	IV	0.73	0.70	0.64	0.67	0.77	0.77	0.69	0.75
13	C12	IV	0.56	0.48	0.50	0.67	0.74	0.74	0.69	0.67

Legitimacy depends on the hydrophobic-hydrophilic properties of the contacting atoms. In order to define it, for each inter-atomic contact, eight atom classes have been introduced:

I Hydrophilic - N and O that can donate and accept hydrogen bonds (e.g., oxygen of hydroxyl group of Ser. or Thr)
II Acceptor - N or O that can only accept a hydrogen bond
III Donor - N that can only donate a hydrogen bond
IV Hydrophobic - Cl, Br, I and all C atoms that are not in aromatic rings and do not have a covalent bond to a N or O atom
V Aromatic - C in aromatic rings irrespective of any other bonds formed by the atom
VI Neutral - C atoms that have a covalent bond to at least one atom of class I or two or more atoms from class II or III; atoms; S, F, P, and metal atoms in all cases
VII Neutral-donor - C atoms that have a covalent bond with only one atom of class III
VIII Neutral-acceptor - C atoms that have a covalent bond with only one atom of class II

For each pair of contacts the state of legitimacy is shown below:

Legend:
+, legitimate
-, illegitimate

Atomic class	I	II	III	IV	V	VI	VII	VIII
I (Hydrophilic)	+	+	+	-	+	+	+	+
II (Acceptor)	+	-	+	-	+	+	+	-
III (Donor)	+	+	-	-	+	+	-	+
IV (Hydrophobic)	-	-	-	+	+	+	+	+
V (Aromatic)	+	+	+	+	+	+	+	+

VI (Neutral) + + + + + + + +
VII (Neutral-donor) + + - + + + - +
VIII (Neutral-acceptor) + - + + + + + -

Contacts of ligand OH 501 in PDB entry 1F91

Table II
Residues in contact with ligand OH 501 in PDB entry 1F91
Legend:
Dist - nearest distance (Å) between atoms of the ligand and the residue
Surf - contact surface area ($Å^2$) between the ligand and the residue
HB - hydrophilic-hydrophilic contact (hydrogen bond)
Arom - aromatic-aromatic contact
Phob - hydrophobic-hydrophobic contact
DC - hydrophobic-hydrophilic contact (destabilizing contact)
+/- - indicates presence/absence of a specific contacts
* - indicates residues contacting ligand by their side chain (including CA atoms)

Residue	Dist	Surf	HB	Arom	Phob	DC
			Specific contacts			
163A SER	4.0	2.1	+	-	-	-
298A HIS*	3.6	5.2	+	-	-	-
328A LYS*	3.4	12.1	+	-	-	-
338A ALA*	3.6	19.9	+	-	-	+
339A GLY	3.9	0.3	+	-	-	-
342A GLU*	2.7	28.2	+	-	-	-
388A ASN	3.5	4.2	+	-	-	-
389A SER*	2.8	33.6	+	-	-	-

Table III
List of putative hydrogen bonds between ligand OH 501 and protein in PDB entry 1F91
Legend:
N - ligand atom number in PDB entry
Dist - distance (Å) between the ligand and protein atoms
Surf - contact surface area ($Å^2$) between the ligand and protein atoms

Ligand atom			Protein atom			Dist	Surf
N	Name	Class	Residue	Name	Class		
1	O	I	GLU 342A	OE1	II	2.7	28.2
1	O	I	SER 389A	OG	I	2.8	27.2
1	O	I	LYS 328A	NZ	III	3.4	12.1
1	O	I	ASN 388A	O	II	3.5	4.2
1	O	I	HIS 298A	ND1	I	3.6	5.2
1	O	I	GLY 339A	N	III	3.9	0.3
1	O	I	ALA 338A	O	II	3.9	0.2
1	O	I	SER 163A	O	II	4.0	2.1

Table IV
Full list of atomic contacts with ligand OH 501 in PDB entry 1F91
Total number of contacts is 11
Legend:
N - ligand atom number in PDB entry
Dist - distance (A) between the ligand and protein atoms

Surf - contact surface area (A**2) between the ligand and protein atoms
* - indicates destabilizing contacts

Ligand atom			Protein atom			Dist	Surf
N	Name	Class	Residue	Name	Class		
1	O	I	GLU 342A	OE1	II	2.7	28.2
1	O	I	SER 389A	OG	I	2.8	27.2
1	O	I	LYS 328A	NZ	III	3.4	12.1
1	O	I	SER 389A	CB	VI	3.4	6.4
1	O	I	ASN 388A	O	II	3.5	4.2
1	O	I	HIS 298A	ND1	I	3.6	5.2
1	O	I	ALA 338A	CB	IV	3.6	16.1*
1	O	I	ALA 338A	C	VI	3.8	3.6
1	O	I	GLY 339A	N	III	3.9	0.3
1	O	I	ALA 338A	O	II	3.9	0.2
1	O	I	SER 163A	O	II	4.0	2.1

Table VI
Normalised complementarity as a function of atomic substitution for ligand OH 501 in PDB entry 1F91
Legend:
N - ligand atom number in PDB entry
Bold - indicates atomic substitution which could stabilize the complex
Italics - indicates atomic substitution which could destabilize the complex

Ligand atom			Atom class							
N	Type	Class	I	II	III	IV	V	VI	VII	VIII
1	O	I	*0.70*	*0.04*	*0.46*	*-0.50*	1.00	1.00	0.76	*0.34*

Legitimacy depends on the hydrophobic-hydrophilic properties of the contacting atoms. In order to define it, for each inter-atomic contact, eight atom classes have been introduced:

I Hydrophilic - N and O that can donate and accept hydrogen bonds (e.g., oxygen of hydroxyl group of Ser. or Thr)
II Acceptor - N or O that can only accept a hydrogen bond
III Donor - N that can only donate a hydrogen bond
IV Hydrophobic - Cl, Br, I and all C atoms that are not in aromatic rings and do not have a covalent bond to a N or O atom
V Aromatic - C in aromatic rings irrespective of any other bonds formed by the atom
VI Neutral - C atoms that have a covalent bond to at least one atom of class I or two or more atoms from class II or III; atoms; S, F, P, and metal atoms in all cases
VII Neutral-donor - C atoms that have a covalent bond with only one atom of class III
VIII Neutral-acceptor - C atoms that have a covalent bond with only one atom of class II

For each pair of contacts the state of legitimacy is shown below:

Legend:

+, legitimate
-, illegitimate

Atomic class	I	II	III	IV	V	VI	VII	VIII
I (Hydrophilic)	+	+	+	-	+	+	+	+
II (Acceptor)	+	-	+	-	+	+	+	-
III (Donor)	+	+	-	-	+	+	-	+
IV (Hydrophobic)	-	-	-	+	+	+	+	+
V (Aromatic)	+	+	+	+	+	+	+	+
VI (Neutral)	+	+	+	+	+	+	+	+
VII (Neutral-donor)	+	+	-	+	+	+	-	+
VIII (Neutral-acceptor)	+	-	+	+	+	+	+	-

Contacts of ligand TLM 500 in PDB entry 1FJ4

Table II
Residues in contact with ligand TLM 500 in PDB entry 1FJ4
Legend:
Dist - nearest distance (Å) between atoms of the ligand and the residue
Surf - contact surface area (Å^2) between the ligand and the residue
HB - hydrophilic-hydrophilic contact (hydrogen bond)
Arom - aromatic-aromatic contact
Phob - hydrophobic-hydrophobic contact
DC - hydrophobic-hydrophilic contact (destabilizing contact)
+/- - indicates presence/absence of a specific contacts
* - indicates residues contacting ligand by their side chain
(including CA atoms)

Residue	Dist	Surf	HB	Arom	Phob	DC
			Specific contacts			
163A CYS*	3.2	13.0	-	-	-	+
206A ALA*	5.5	1.8	-	-	+	-
229A PHE*	3.4	40.2	-	-	+	-
265A ASP*	5.2	0.4	-	-	-	+
268A ASP	3.8	5.2	-	-	-	+
270A VAL*	3.5	13.0	+	-	-	+
271A ALA	3.3	41.3	-	-	-	+
272A PRO*	3.9	30.3	-	-	+	+
298A HIS*	2.8	21.7	+	-	+	-
300A THR*	3.7	20.9	-	-	+	+
302A THR*	4.0	14.4	+	-	+	+
305A GLY*	4.0	27.0	+	-	-	-
333A HIS*	2.6	31.2	+	-	+	+
390A PHE*	3.3	46.4	-	-	+	+
391A GLY*	3.5	18.8	-	-	-	+
392A PHE*	3.7	30.1	-	-	+	-
393A GLY	3.8	4.5	-	-	-	+
394A GLY*	3.3	35.2	-	-	-	+
396A ASN*	5.0	5.6	-	-	-	+

Table III
List of putative hydrogen bonds between ligand TLM 500 and protein in PDB entry 1FJ4
Legend:
N - ligand atom number in PDB entry
Dist - distance (Å) between the ligand and protein atoms
Surf - contact surface area (Å^2) between the ligand and protein atoms

Ligand atom			Protein atom			Dist	Surf
N	Name	Class	Residue	Name	Class		

N	Name	Class	Residue		Name	Class	Dist	Surf
10	O1	II	HIS	333A	NE2	I	2.6	29.6
10	O1	II	HIS	298A	NE2	I	2.8	5.5
12	O2	I	THR	302A	OG1	I	4.0	9.7
12	O2	I	VAL	270A	O	II	4.4	4.5
12	O2	I	GLY	305A	N	III	5.4	0.2

Table IV

Full list of atomic contacts with ligand TLM 500 in PDB entry 1FJ4

Total number of contacts is 81

Legend:

N - ligand atom number in PDB entry

Dist - distance (A) between the ligand and protein atoms

Surf - contact surface area (A**2) between the ligand and protein atoms

* - indicates destabilizing contacts

Ligand atom			Protein atom				Dist	Surf
N	Name	Class	Residue		Name	Class		
1	S1	VI	PHE	390A	O	II	3.3	20.9
1	S1	VI	HIS	298A	NE2	I	3.5	5.2
1	S1	VI	PHE	390A	C	VI	3.6	1.6
1	S1	VI	HIS	298A	CD2	V	3.7	1.1
1	S1	VI	GLY	391A	N	III	3.9	0.9
1	S1	VI	GLY	391A	CA	VI	3.9	2.2
1	S1	VI	GLY	391A	C	VI	4.4	0.2
2	C1	VIII	HIS	298A	NE2	I	3.3	4.5
2	C1	VIII	THR	300A	CG2	IV	3.7	1.6
2	C1	VIII	PHE	390A	O	II	4.0	0.2*
2	C1	VIII	CYS	163A	SG	VI	4.1	4.9
2	C1	VIII	PHE	392A	CE1	V	4.4	1.3
3	C2	IV	PHE	392A	CE1	V	3.8	2.5
3	C2	IV	THR	300A	CG2	IV	3.9	2.2
4	C3	VI	PHE	392A	CE1	V	3.7	2.5
4	C3	VI	THR	300A	OG1	I	4.3	1.6
6	C5	IV	PHE	392A	CD1	V	3.8	7.2
6	C5	IV	PHE	392A	CE1	V	3.9	0.7
6	C5	IV	PRO	272A	CD	IV	4.0	4.5
6	C5	IV	VAL	270A	O	II	4.3	1.8*
7	C6	IV	GLY	391A	O	II	3.9	1.6*
7	C6	IV	PRO	272A	CD	IV	3.9	2.2
8	C7	IV	ALA	271A	N	III	3.3	16.6*
8	C7	IV	VAL	270A	C	VI	3.5	3.1
8	C7	IV	ALA	271A	C	VI	3.5	2.0
8	C7	IV	VAL	270A	CA	VII	3.8	2.9
8	C7	IV	GLY	391A	O	II	3.8	2.2*
8	C7	IV	VAL	270A	N	III	3.9	0.2*
8	C7	IV	PHE	392A	CA	VII	3.9	2.2
9	C8	IV	GLY	394A	N	III	3.3	29.8*
9	C8	IV	ALA	271A	O	II	3.5	18.4*
9	C8	IV	ALA	271A	C	VI	3.6	0.9
9	C8	IV	ALA	271A	N	III	3.6	3.4*
9	C8	IV	GLY	391A	O	II	3.6	1.6*
9	C8	IV	GLY	394A	CA	VI	3.7	4.7
9	C8	IV	ASP	268A	O	II	3.8	5.2*
9	C8	IV	GLY	393A	N	III	3.8	4.5*

9	C8	IV	PRO	272A	N	VI	3.9	0.2
9	C8	IV	VAL	270A	N	III	3.9	0.4*
9	C8	IV	PRO	272A	CA	IV	4.2	1.1
10	O1	II	HIS	333A	NE2	I	2.6	29.6
10	O1	II	HIS	298A	NE2	I	2.8	5.5
10	O1	II	CYS	163A	CB	IV	3.2	6.6*
10	O1	II	CYS	163A	SG	VI	3.2	0.3
11	C9	IV	PHE	229A	CD2	V	3.4	32.5
11	C9	IV	PHE	229A	CE2	V	3.5	7.6
11	C9	IV	THR	300A	OG1	I	4.0	9.2*
11	C9	IV	THR	300A	CG2	IV	4.0	2.0
11	C9	IV	HIS	333A	CD2	V	4.1	1.3
11	C9	IV	PHE	392A	CE1	V	4.2	7.6
11	C9	IV	THR	302A	CG2	IV	4.2	2.9
11	C9	IV	HIS	333A	NE2	I	4.2	0.2*
11	C9	IV	PHE	392A	CZ	V	4.4	0.9
11	C9	IV	CYS	163A	SG	VI	4.9	1.1
11	C9	IV	ALA	206A	CB	IV	5.5	1.8
12	O2	I	PHE	392A	CE1	V	3.8	5.0
12	O2	I	THR	302A	OG1	I	4.0	9.7
12	O2	I	THR	302A	CG2	IV	4.3	1.2*
12	O2	I	VAL	270A	O	II	4.4	4.5
12	O2	I	PHE	392A	CZ	V	4.5	0.2
12	O2	I	THR	302A	CB	VI	4.6	0.5
12	O2	I	GLY	305A	CA	VI	5.2	0.2
12	O2	I	PRO	272A	CD	IV	5.3	0.9*
12	O2	I	GLY	305A	N	III	5.4	0.2
13	C10	IV	GLY	305A	CA	VI	4.0	26.0
13	C10	IV	PRO	272A	CG	IV	4.0	8.1
13	C10	IV	PRO	272A	CB	IV	4.6	0.4
13	C10	IV	THR	300A	OG1	I	4.7	4.0*
13	C10	IV	HIS	298A	CD2	V	4.7	5.4
13	C10	IV	GLY	305A	C	VI	4.9	0.7
13	C10	IV	THR	300A	CG2	IV	5.0	0.2
13	C10	IV	PHE	390A	CB	IV	5.1	1.3
13	C10	IV	PHE	390A	CD2	V	5.2	3.1
14	C11	IV	GLY	391A	O	II	3.5	10.1*
14	C11	IV	PHE	390A	CB	IV	3.6	19.1
14	C11	IV	PRO	272A	CB	IV	3.9	12.6
14	C11	IV	PRO	272A	CA	IV	4.3	0.2
14	C11	IV	PHE	390A	CG	V	4.6	0.2
14	C11	IV	ASN	396A	OD1	II	5.0	5.6*
14	C11	IV	GLY	394A	CA	VI	5.1	0.7
14	C11	IV	ASP	265A	OD2	II	5.2	0.4*

Table VI
Normalised complementarity as a function of atomic substitution for ligand TLM 500 in PDB entry 1FJ4

Legend:

N - ligand atom number in PDB entry

Bold - indicates atomic substitution which could stabilize the complex

Italics - indicates atomic substitution which could destabilize the complex

Ligand atom	Atom class

N	Type	Class	I	II	III	IV	V	VI	VII	VIII
1	S1	VI	0.38	0.27	0.37	0.24	0.38	0.38	0.37	0.27
2	C1	VIII	0.37	0.37	0.37	0.35	0.38	0.38	0.38	0.38
3	C2	IV	0.36	0.36	0.36	0.38	0.38	0.38	0.38	0.38
4	C3	VI	0.38	0.38	0.38	0.37	0.38	0.38	0.38	0.38
6	C5	IV	0.36	0.35	0.36	0.38	0.38	0.38	0.38	0.38
7	C6	IV	0.37	0.36	0.37	0.38	0.38	0.38	0.38	0.38
8	C7	IV	0.47	0.46	0.36	0.38	0.47	0.47	0.36	0.46
9	C8	IV	0.68	0.56	0.49	0.38	0.69	0.69	0.50	0.56
10	O1	II	0.38	0.38	0.38	0.24	0.41	0.41	0.41	0.41
11	C9	IV	0.39	0.39	0.39	0.38	0.42	0.42	0.42	0.42
12	O2	I	0.38	0.35	0.38	0.32	0.39	0.39	0.39	0.36
13	C10	IV	0.35	0.35	0.35	0.38	0.40	0.40	0.40	0.40
14	C11	IV	0.30	0.22	0.30	0.38	0.46	0.46	0.46	0.38

Legitimacy depends on the hydrophobic-hydrophilic properties of the contacting atoms. In order to define it, for each inter-atomic contact, eight atom classes have been introduced:

I Hydrophilic - N and O that can donate and accept hydrogen bonds (e.g., oxygen of hydroxyl group of Ser. or Thr)
II Acceptor - N or O that can only accept a hydrogen bond
III Donor - N that can only donate a hydrogen bond
IV Hydrophobic - Cl, Br, I and all C atoms that are not in aromatic rings and do not have a covalent bond to a N or O atom
V Aromatic - C in aromatic rings irrespective of any other bonds formed by the atom
VI Neutral - C atoms that have a covalent bond to at least one atom of class I or two or more atoms from class II or III; atoms; S, F, P, and metal atoms in all cases
VII Neutral-donor - C atoms that have a covalent bond with only one atom of class III
VIII Neutral-acceptor - C atoms that have a covalent bond with only one atom of class II

For each pair of contacts the state of legitimacy is shown below:

Legend:
+, legitimate
-, illegitimate

Atomic class	I	II	III	IV	V	VI	VII	VIII
I (Hydrophilic)	+	+	+	-	+	+	+	+
II (Acceptor)	+	-	+	-	+	+	+	-

III (Donor)	+	+	-	-	+	+	-	+
IV (Hydrophobic)	-	-	-	+	+	+	+	+
V (Aromatic)	+	+	+	+	+	+	+	+
VI (Neutral)	+	+	+	+	+	+	+	+
VII (Neutral-donor)	+	+	-	+	+	+	-	+
VIII (Neutral-acceptor)	+	-	+	+	+	+	+	-

Contacts of ligand TLM 501 in PDB entry 1FJ4

Table II
Residues in contact with ligand TLM 501 in PDB entry 1FJ4
Legend:
Dist - nearest distance (Å) between atoms of the ligand and the residue
Surf - contact surface area (Å^2) between the ligand and the residue
HB - hydrophilic-hydrophilic contact (hydrogen bond)
Arom - aromatic-aromatic contact
Phob - hydrophobic-hydrophobic contact
DC - hydrophobic-hydrophilic contact (destabilizing contact)
+/- - indicates presence/absence of a specific contacts
* - indicates residues contacting ligand by their side chain
(including CA atoms)

Residue	Dist	Surf	Specific contacts			
			HB	Arom	Phob	DC
163B CYS*	3.2	12.4	-	-	-	-
206B ALA*	5.6	1.8	-	-	+	-
229B PHE*	3.4	39.7	-	-	+	-
265B ASP*	5.2	0.2	-	-	-	+
268B ASP	3.8	6.1	-	-	-	+
270B VAL*	3.4	12.8	+	-	-	+
271B ALA	3.2	42.2	-	-	-	+
272B PRO*	3.9	30.3	-	-	+	+
298B HIS*	2.9	23.6	+	-	+	-
300B THR*	3.7	20.0	-	-	+	+
302B THR*	3.9	16.6	+	-	+	+
305B GLY*	3.9	26.6	+	-	-	-
333B HIS*	2.7	30.6	+	-	+	+
390B PHE*	3.3	44.4	-	-	+	+
391B GLY*	3.6	19.3	-	-	-	+
392B PHE*	3.8	29.7	-	-	+	-
393B GLY	3.8	5.4	-	-	-	+
394B GLY*	3.4	33.0	-	-	-	+
396B ASN*	5.0	5.6	-	-	-	+

Table III
List of putative hydrogen bonds between ligand TLM 501 and protein in PDB entry 1FJ4
Legend:
N - ligand atom number in PDB entry
Dist - distance (Å) between the ligand and protein atoms
Surf - contact surface area (Å^2) between the ligand and protein atoms

Ligand atom			Protein atom			Dist	Surf
N	Name	Class	Residue	Name	Class		
10	O1	II	HIS 333B	NE2	I	2.7	30.0

10	O1	II	HIS 298B	NE2	I	2.9	5.9
12	O2	I	THR 302B	OG1	I	3.9	11.1
12	O2	I	VAL 270B	O	II	4.5	4.7
12	O2	I	GLY 305B	N	III	5.2	0.3

Table IV

Full list of atomic contacts with ligand TLM 501 in PDB entry 1FJ4

Total number of contacts is 80

Legend:

N - ligand atom number in PDB entry

Dist - distance (A) between the ligand and protein atoms

Surf - contact surface area (A**2) between the ligand and protein atoms

* - indicates destabilizing contacts

Ligand atom			Protein atom			Dist	Surf
N	Name	Class	Residue	Name	Class		
1	S1	VI	PHE 390B	O	II	3.3	20.6
1	S1	VI	HIS 298B	NE2	I	3.5	4.7
1	S1	VI	PHE 390B	C	VI	3.6	1.3
1	S1	VI	HIS 298B	CD2	V	3.8	1.1
1	S1	VI	GLY 391B	N	III	3.9	0.9
1	S1	VI	GLY 391B	CA	VI	3.9	3.1
1	S1	VI	PHE 392B	CE1	V	4.7	0.2
2	C1	VIII	HIS 298B	NE2	I	3.4	4.5
2	C1	VIII	THR 300B	CG2	IV	3.7	1.3
2	C1	VIII	PHE 390B	O	II	4.1	0.2*
2	C1	VIII	CYS 163B	SG	VI	4.1	5.2
2	C1	VIII	PHE 392B	CE1	V	4.4	1.3
3	C2	IV	THR 300B	CG2	IV	3.9	2.2
3	C2	IV	PHE 392B	CE1	V	3.9	2.7
4	C3	VI	PHE 392B	CE1	V	3.8	2.2
4	C3	VI	THR 300B	OG1	I	4.3	1.6
6	C5	IV	PHE 392B	CD1	V	3.9	6.1
6	C5	IV	PRO 272B	CD	IV	3.9	4.7
6	C5	IV	PHE 392B	CE1	V	3.9	0.7
6	C5	IV	VAL 270B	O	II	4.2	1.6*
7	C6	IV	PRO 272B	CD	IV	3.9	2.7
7	C6	IV	GLY 391B	O	II	4.0	1.3*
7	C6	IV	GLY 391B	C	VI	4.1	0.2
8	C7	IV	ALA 271B	N	III	3.2	17.7*
8	C7	IV	VAL 270B	C	VI	3.4	2.9
8	C7	IV	ALA 271B	C	VI	3.5	1.8
8	C7	IV	VAL 270B	CA	VII	3.8	2.9
8	C7	IV	PHE 392B	CA	VII	3.9	4.7
9	C8	IV	GLY 394B	N	III	3.4	28.3*
9	C8	IV	ALA 271B	O	II	3.5	18.4*
9	C8	IV	ALA 271B	N	III	3.5	4.0*
9	C8	IV	ALA 271B	C	VI	3.6	0.2
9	C8	IV	GLY 391B	O	II	3.7	2.2*
9	C8	IV	ASP 268B	O	II	3.8	6.1*
9	C8	IV	GLY 394B	CA	VI	3.8	3.6
9	C8	IV	GLY 393B	N	III	3.8	5.4*

9	C8	IV	VAL	270B	N	III	3.8	0.7*
9	C8	IV	PRO	272B	N	VI	3.9	0.2
9	C8	IV	PRO	272B	CA	IV	4.2	1.1
10	O1	II	HIS	333B	NE2	I	2.7	30.0
10	O1	II	HIS	298B	NE2	I	2.9	5.9
10	O1	II	CYS	163B	SG	VI	3.2	6.6
11	C9	IV	PHE	229B	CD2	V	3.4	33.0
11	C9	IV	PHE	229B	CE2	V	3.6	6.7
11	C9	IV	THR	300B	OG1	I	3.8	10.3*
11	C9	IV	THR	300B	CB	VI	3.9	1.6
11	C9	IV	THR	300B	CG2	IV	4.0	0.2
11	C9	IV	THR	302B	CG2	IV	4.0	3.6
11	C9	IV	HIS	333B	CD2	V	4.1	0.4
11	C9	IV	PHE	392B	CE1	V	4.3	6.5
11	C9	IV	HIS	333B	NE2	I	4.4	0.2*
11	C9	IV	PHE	392B	CZ	V	4.5	1.1
11	C9	IV	CYS	163B	SG	VI	5.1	0.7
11	C9	IV	ALA	206B	CB	IV	5.6	1.8
12	O2	I	THR	302B	OG1	I	3.9	11.1
12	O2	I	PHE	392B	CE1	V	4.0	4.2
12	O2	I	THR	302B	CG2	IV	4.2	1.4*
12	O2	I	VAL	270B	O	II	4.5	4.7
12	O2	I	THR	302B	CB	VI	4.5	0.5
12	O2	I	GLY	305B	CA	VI	5.0	0.7
12	O2	I	PRO	272B	CD	IV	5.2	0.9*
12	O2	I	GLY	305B	N	III	5.2	0.3
13	C10	IV	GLY	305B	CA	VI	3.9	25.4
13	C10	IV	PRO	272B	CG	IV	4.0	7.2
13	C10	IV	PRO	272B	CD	IV	4.2	0.2
13	C10	IV	PRO	272B	CB	IV	4.5	0.2
13	C10	IV	HIS	298B	CD2	V	4.6	7.4
13	C10	IV	THR	300B	OG1	I	4.7	2.7*
13	C10	IV	GLY	305B	C	VI	4.9	0.2
13	C10	IV	PHE	390B	CB	IV	5.0	1.6
13	C10	IV	PHE	390B	CD2	V	5.2	3.4
14	C11	IV	GLY	391B	O	II	3.6	11.2*
14	C11	IV	PHE	390B	CB	IV	3.6	17.0
14	C11	IV	GLY	391B	N	III	3.9	0.2*
14	C11	IV	PRO	272B	CB	IV	3.9	12.8
14	C11	IV	PRO	272B	CA	IV	4.3	0.2
14	C11	IV	PHE	390B	CG	V	4.7	0.2
14	C11	IV	ASN	396B	OD1	II	5.0	5.6*
14	C11	IV	GLY	394B	CA	VI	5.1	1.1
14	C11	IV	ASP	265B	OD2	II	5.2	0.2*

Table VI

Normalised complementarity as a function of atomic substitution for ligand TLM 501 in PDB entry 1FJ4

Legend:

N - ligand atom number in PDB entry

Bold - indicates atomic substitution which could stabilize the complex

Italics - indicates atomic substitution which could destabilize the complex

Ligand atom			Atom class							
N	Type	Class	I	II	III	IV	V	VI	VII	VIII

1	S1	VI	0.40	0.30	0.40	0.27	0.40	0.40	0.40	0.30
2	C1	VIII	0.39	0.39	0.39	0.38	0.40	0.40	0.40	0.40
3	C2	IV	0.39	0.39	0.39	0.40	0.40	0.40	0.40	0.40
4	C3	VI	0.40	0.40	0.40	0.39	0.40	0.40	0.40	0.40
6	C5	IV	0.38	0.38	0.38	0.40	0.41	0.41	0.41	0.40
7	C6	IV	0.39	0.39	0.39	0.40	0.41	0.41	0.41	0.40
8	C7	IV	0.49	0.49	0.36	0.40	0.49	0.49	0.36	0.49
9	C8	IV	0.71	0.58	0.53	0.40	0.72	0.72	0.53	0.59
10	O1	II	0.40	0.40	0.40	0.22	0.40	0.40	0.40	0.40
11	C9	IV	0.42	0.42	0.42	0.40	0.45	0.45	0.45	0.45
12	O2	I	0.40	0.38	0.40	0.33	0.41	0.41	0.41	0.39
13	C10	IV	0.37	0.37	0.37	0.40	0.41	0.41	0.41	0.41
14	C11	IV	0.34	0.25	0.34	0.40	0.48	0.48	0.48	0.40

Legitimacy depends on the hydrophobic-hydrophilic properties of the contacting atoms. In order to define it, for each inter-atomic contact, eight atom classes have been introduced:

I Hydrophilic - N and O that can donate and accept hydrogen bonds (e.g., oxygen of hydroxyl group of Ser. or Thr)
II Acceptor - N or O that can only accept a hydrogen bond
III Donor - N that can only donate a hydrogen bond
IV Hydrophobic - Cl, Br, I and all C atoms that are not in aromatic rings and do not have a covalent bond to a N or O atom
V Aromatic - C in aromatic rings irrespective of any other bonds formed by the atom
VI Neutral - C atoms that have a covalent bond to at least one atom of class I or two or more atoms from class II or III; atoms; S, F, P, and metal atoms in all cases
VII Neutral-donor - C atoms that have a covalent bond with only one atom of class III
VIII Neutral-acceptor - C atoms that have a covalent bond with only one atom of class II

For each pair of contacts the state of legitimacy is shown below:

Legend:
+, legitimate
-, illegitimate

Atomic class	I	II	III	IV	V	VI	VII	VIII
I (Hydrophilic)	+	+	+	-	+	+	+	+
II (Acceptor)	+	-	+	-	+	+	+	-
III (Donor)	+	+	-	-	+	+	-	+
IV (Hydrophobic)	-	-	-	+	+	+	+	+

V (Aromatic)	+	+	+	+	+	+	+	+
VI (Neutral)	+	+	+	+	+	+	+	+
VII (Neutral-donor)	+	+	-	+	+	+	-	+
VIII (Neutral-acceptor)	+	-	+	+	+	+	+	-

Contacts of ligand TLM 502 in PDB entry 1FJ4

Table II
Residues in contact with ligand TLM 502 in PDB entry 1FJ4
Legend:
Dist - nearest distance (Å) between atoms of the ligand and the residue
Surf - contact surface area ($Å^2$) between the ligand and the residue
HB - hydrophilic-hydrophilic contact (hydrogen bond)
Arom - aromatic-aromatic contact
Phob - hydrophobic-hydrophobic contact
DC - hydrophobic-hydrophilic contact (destabilizing contact)
+/- - indicates presence/absence of a specific contacts
* - indicates residues contacting ligand by their side chain
(including CA atoms)

Residue	Dist	Surf	HB	Arom	Phob	DC
			Specific contacts			
163C CYS*	3.2	13.0	-	-	-	-
206C ALA*	5.4	2.0	-	-	+	-
229C PHE*	3.5	39.7	-	-	+	-
265C ASP*	5.1	0.4	-	-	-	+
268C ASP	3.7	6.7	-	-	-	+
270C VAL*	3.5	13.7	+	-	-	+
271C ALA	3.2	41.3	-	-	-	+
272C PRO*	3.8	40.1	-	-	+	+
298C HIS*	3.0	21.6	+	-	+	-
300C THR*	3.8	19.3	-	-	+	+
302C THR*	4.0	14.9	+	-	+	+
305C GLY*	4.0	17.3	+	-	-	-
333C HIS*	2.7	30.9	+	-	+	-
390C PHE*	3.3	43.5	-	-	+	-
391C GLY*	3.6	19.1	-	-	-	+
392C PHE*	3.6	31.0	-	-	+	-
393C GLY	3.9	3.8	-	-	-	+
394C GLY*	3.3	34.1	-	-	-	+
396C ASN*	5.0	5.6	-	-	-	+

Table III
List of putative hydrogen bonds between ligand TLM 502 and protein in PDB entry 1FJ4
Legend:
N - ligand atom number in PDB entry
Dist - distance (Å) between the ligand and protein atoms
Surf - contact surface area ($Å^2$) between the ligand and protein atoms

Ligand atom			Protein atom			Dist	Surf
N	Name	Class	Residue	Name	Class		

10	O1	II	HIS 333C	NE2	I	2.7	30.0
10	O1	II	HIS 298C	NE2	I	3.0	5.0
12	O2	I	THR 302C	OG1	I	4.0	10.6
12	O2	I	VAL 270C	O	II	4.3	5.2
12	O2	I	GLY 305C	N	III	5.3	0.3

Table IV
Full list of atomic contacts with ligand TLM 502 in PDB entry 1FJ4
Total number of contacts is 79
Legend:
N - ligand atom number in PDB entry
Dist - distance (A) between the ligand and protein atoms
Surf - contact surface area (A**2) between the ligand and protein atoms
* - indicates destabilizing contacts

Ligand atom			Protein atom				
N	Name	Class	Residue	Name	Class	Dist	Surf
1	S1	VI	PHE 390C	O	II	3.3	19.7
1	S1	VI	HIS 298C	NE2	I	3.6	5.8
1	S1	VI	PHE 390C	C	VI	3.6	2.2
1	S1	VI	HIS 298C	CD2	V	3.8	0.9
1	S1	VI	GLY 391C	N	III	3.9	1.3
1	S1	VI	GLY 391C	CA	VI	3.9	2.0
1	S1	VI	GLY 391C	C	VI	4.3	0.2
2	C1	VIII	HIS 298C	NE2	I	3.5	4.3
2	C1	VIII	THR 300C	CG2	IV	3.8	1.8
2	C1	VIII	CYS 163C	SG	VI	4.1	4.7
2	C1	VIII	PHE 392C	CE1	V	4.3	1.8
3	C2	IV	PHE 392C	CE1	V	3.7	2.7
3	C2	IV	THR 300C	CG2	IV	4.1	1.8
4	C3	VI	PHE 392C	CE1	V	3.6	2.7
4	C3	VI	THR 300C	OG1	I	4.4	1.3
6	C5	IV	PHE 392C	CD1	V	3.8	6.1
6	C5	IV	PHE 392C	CE1	V	3.9	0.4
6	C5	IV	PRO 272C	CD	IV	3.9	4.7
6	C5	IV	VAL 270C	O	II	4.2	1.6*
7	C6	IV	PRO 272C	N	VI	3.9	2.0
7	C6	IV	GLY 391C	O	II	3.9	1.6*
7	C6	IV	GLY 391C	C	VI	4.1	0.2
8	C7	IV	ALA 271C	N	III	3.2	17.9*
8	C7	IV	ALA 271C	C	VI	3.4	1.1
8	C7	IV	VAL 270C	C	VI	3.5	2.9
8	C7	IV	VAL 270C	CA	VII	3.8	3.1
8	C7	IV	VAL 270C	N	III	3.8	0.4*
8	C7	IV	GLY 391C	O	II	3.9	2.5*
8	C7	IV	PHE 392C	CA	VII	4.0	2.2
9	C8	IV	GLY 394C	N	III	3.3	29.2*
9	C8	IV	ALA 271C	O	II	3.4	18.4*
9	C8	IV	ALA 271C	C	VI	3.5	0.4
9	C8	IV	ALA 271C	N	III	3.5	3.4*
9	C8	IV	ASP 268C	O	II	3.7	6.7*
9	C8	IV	GLY 394C	CA	VI	3.7	3.4
9	C8	IV	GLY 391C	O	II	3.8	1.6*

9	C8	IV	PRO	272C	N	VI	3.8	0.4
9	C8	IV	GLY	393C	N	III	3.9	3.8*
9	C8	IV	VAL	270C	N	III	3.9	0.4*
9	C8	IV	PRO	272C	CA	IV	4.1	1.8
10	O1	II	HIS	333C	NE2	I	2.7	30.0
10	O1	II	HIS	298C	NE2	I	3.0	5.0
10	O1	II	CYS	163C	SG	VI	3.2	6.9
11	C9	IV	PHE	229C	CD2	V	3.5	32.8
11	C9	IV	PHE	229C	CE2	V	3.6	7.0
11	C9	IV	PHE	392C	CE1	V	4.0	9.2
11	C9	IV	THR	300C	OG1	I	4.1	9.0*
11	C9	IV	THR	300C	CG2	IV	4.2	1.6
11	C9	IV	HIS	333C	CD2	V	4.2	0.9
11	C9	IV	THR	302C	CG2	IV	4.2	2.9
11	C9	IV	PHE	392C	CZ	V	4.2	1.6
11	C9	IV	CYS	163C	SG	VI	5.0	1.3
11	C9	IV	ALA	206C	CB	IV	5.4	2.0
12	O2	I	PHE	392C	CE1	V	3.8	4.2
12	O2	I	THR	302C	OG1	I	4.0	10.6
12	O2	I	VAL	270C	O	II	4.3	5.2
12	O2	I	THR	302C	CG2	IV	4.4	1.0*
12	O2	I	PHE	392C	CZ	V	4.6	0.2
12	O2	I	THR	302C	CB	VI	4.7	0.3
12	O2	I	GLY	305C	CA	VI	5.1	0.3
12	O2	I	PRO	272C	CD	IV	5.1	0.9*
12	O2	I	GLY	305C	N	III	5.3	0.3
13	C10	IV	PRO	272C	CG	IV	3.9	16.6
13	C10	IV	GLY	305C	CA	VI	4.0	16.4
13	C10	IV	PRO	272C	CD	IV	4.1	0.2
13	C10	IV	PRO	272C	CB	IV	4.5	0.4
13	C10	IV	THR	300C	OG1	I	4.7	3.6*
13	C10	IV	HIS	298C	CD2	V	4.7	5.6
13	C10	IV	GLY	305C	C	VI	4.9	0.2
13	C10	IV	PHE	390C	CB	IV	5.0	1.6
13	C10	IV	THR	300C	CG2	IV	5.1	0.2
13	C10	IV	PHE	390C	CD2	V	5.3	2.7
14	C11	IV	GLY	391C	O	II	3.6	9.6*
14	C11	IV	PHE	390C	CB	IV	3.7	17.0
14	C11	IV	PRO	272C	CB	IV	3.8	13.0
14	C11	IV	PHE	390C	CG	V	4.7	0.2
14	C11	IV	ASN	396C	OD1	II	5.0	5.6*
14	C11	IV	GLY	394C	CA	VI	5.0	1.6
14	C11	IV	ASP	265C	OD2	II	5.1	0.4*

Table VI
Normalised complementarity as a function of atomic substitution for ligand TLM 502 in PDB entry 1FJ4

Legend:

N - ligand atom number in PDB entry

Bold - indicates atomic substitution which could stabilize the complex

Italics - indicates atomic substitution which could destabilize the complex

Ligand atom			Atom class							
N	Type	Class	I	II	III	IV	V	VI	VII	VIII

1	S1	VI	*0.40*	*0.30*	*0.39*	*0.27*	*0.40*	*0.40*	*0.39*	*0.30*
2	C1	VIII	*0.39*	*0.39*	*0.39*	*0.38*	*0.40*	*0.40*	*0.40*	*0.40*
3	C2	IV	*0.39*	*0.39*	*0.39*	*0.40*	*0.40*	*0.40*	*0.40*	*0.40*
4	C3	VI	*0.40*	*0.40*	*0.40*	*0.39*	*0.40*	*0.40*	*0.40*	*0.40*
6	C5	IV	*0.39*	*0.38*	*0.39*	*0.40*	*0.41*	*0.41*	*0.41*	*0.40*
7	C6	IV	*0.41*	*0.40*	*0.41*	*0.40*	*0.41*	*0.41*	*0.41*	*0.40*
8	C7	IV	0.50	0.49	*0.39*	*0.40*	0.50	0.50	*0.39*	0.49
9	C8	IV	0.71	0.57	0.52	*0.40*	0.71	0.71	0.53	0.58
10	O1	II	*0.40*	*0.40*	*0.40*	*0.23*	*0.40*	*0.40*	*0.40*	*0.40*
11	C9	IV	*0.41*	*0.41*	*0.41*	*0.40*	0.45	0.45	0.45	0.45
12	O2	I	*0.40*	*0.38*	*0.40*	*0.33*	*0.41*	*0.41*	*0.41*	*0.39*
13	C10	IV	*0.33*	*0.33*	*0.33*	*0.40*	*0.42*	*0.42*	*0.42*	*0.42*
14	C11	IV	*0.33*	*0.25*	*0.33*	*0.40*	0.48	0.48	0.48	*0.40*

Legitimacy depends on the hydrophobic-hydrophilic properties of the contacting atoms. In order to define it, for each inter-atomic contact, eight atom classes have been introduced:

I Hydrophilic - N and O that can donate and accept hydrogen bonds (e.g., oxygen of hydroxyl group of Ser. or Thr)
II Acceptor - N or O that can only accept a hydrogen bond
III Donor - N that can only donate a hydrogen bond
IV Hydrophobic - Cl, Br, I and all C atoms that are not in aromatic rings and do not have a covalent bond to a N or O atom
V Aromatic - C in aromatic rings irrespective of any other bonds formed by the atom
VI Neutral - C atoms that have a covalent bond to at least one atom of class I or two or more atoms from class II or III; atoms; S, F, P, and metal atoms in all cases
VII Neutral-donor - C atoms that have a covalent bond with only one atom of class III
VIII Neutral-acceptor - C atoms that have a covalent bond with only one atom of class II

For each pair of contacts the state of legitimacy is shown below:

Legend:
+, legitimate
-, illegitimate

Atomic class	I	II	III	IV	V	VI	VII	VIII
I (Hydrophilic)	+	+	+	-	+	+	+	+
II (Acceptor)	+	-	+	-	+	+	+	-
III (Donor)	+	+	-	-	+	+	-	+
IV (Hydrophobic)	-	-	-	+	+	+	+	+

V (Aromatic)	+	+	+	+	+	+	+	+
VI (Neutral)	+	+	+	+	+	+	+	+
VII (Neutral-donor)	+	+	-	+	+	+	-	+
VIII (Neutral-acceptor)	+	-	+	+	+	+	+	-

Contacts of ligand TLM 503 in PDB entry 1FJ4

Table II
Residues in contact with ligand TLM 503 in PDB entry 1FJ4
Legend:
Dist - nearest distance (Å) between atoms of the ligand and the residue
Surf - contact surface area ($Å^2$) between the ligand and the residue
HB - hydrophilic-hydrophilic contact (hydrogen bond)
Arom - aromatic-aromatic contact
Phob - hydrophobic-hydrophobic contact
DC - hydrophobic-hydrophilic contact (destabilizing contact)
+/- - indicates presence/absence of a specific contacts
* - indicates residues contacting ligand by their side chain (including CA atoms)

Residue	Dist	Surf	HB	Arom	Phob	DC
			Specific contacts			
163D CYS*	3.3	12.5	-	-	-	-
206D ALA*	5.6	1.8	-	-	+	-
229D PHE*	3.5	37.9	-	-	+	-
265D ASP*	4.4	0.9	-	-	-	+
268D ASP	3.7	5.2	-	-	-	+
270D VAL*	3.6	12.3	+	-	-	+
271D ALA	3.3	42.6	-	-	-	+
272D PRO*	3.8	40.0	-	-	+	+
298D HIS*	2.9	23.0	+	-	+	-
300D THR*	3.7	20.6	-	-	+	+
302D THR*	3.9	16.6	+	-	+	+
305D GLY*	4.0	17.4	+	-	-	+
333D HIS*	2.7	30.4	+	-	+	-
390D PHE*	3.2	49.1	-	-	+	-
391D GLY*	3.5	18.2	-	-	-	+
392D PHE*	3.9	28.7	-	-	+	-
393D GLY	3.9	4.5	-	-	-	+
394D GLY*	3.3	34.1	-	-	-	+
396D ASN*	4.8	6.1	-	-	-	+

Table III
List of putative hydrogen bonds between ligand TLM 503 and protein in PDB entry 1FJ4
Legend:
N - ligand atom number in PDB entry
Dist - distance (Å) between the ligand and protein atoms
Surf - contact surface area ($Å^2$) between the ligand and protein atoms

Ligand atom			Protein atom			Dist	Surf
N	Name	Class	Residue	Name	Class		
10	O1	II	HIS 333D	NE2	I	2.7	30.1

N	Name	Class	Residue	Name	Class	Dist	Surf
10	O1	II	HIS 298D	NE2	I	2.9	5.0
12	O2	I	THR 302D	OG1	I	3.9	11.6
12	O2	I	VAL 270D	O	II	4.5	4.9
12	O2	I	GLY 305D	N	III	5.1	0.3

Table IV
Full list of atomic contacts with ligand TLM 503 in PDB entry 1FJ4
Total number of contacts is 79
Legend:
N - ligand atom number in PDB entry
Dist - distance (A) between the ligand and protein atoms
Surf - contact surface area (A**2) between the ligand and protein atoms
* - indicates destabilizing contacts

Ligand atom			Protein atom			Dist	Surf
N	Name	Class	Residue	Name	Class		
1	S1	VI	PHE 390D	O	II	3.2	20.4
1	S1	VI	HIS 298D	NE2	I	3.5	4.7
1	S1	VI	PHE 390D	C	VI	3.5	2.0
1	S1	VI	HIS 298D	CD2	V	3.7	0.7
1	S1	VI	GLY 391D	N	III	3.8	0.9
1	S1	VI	GLY 391D	CA	VI	3.9	2.5
1	S1	VI	GLY 391D	C	VI	4.3	0.2
2	C1	VIII	HIS 298D	NE2	I	3.4	4.7
2	C1	VIII	THR 300D	CG2	IV	3.7	1.6
2	C1	VIII	CYS 163D	CB	IV	4.2	2.2
2	C1	VIII	CYS 163D	SG	VI	4.2	2.2
2	C1	VIII	PHE 392D	CE1	V	4.4	2.2
3	C2	IV	THR 300D	CG2	IV	3.9	2.5
3	C2	IV	PHE 392D	CE1	V	3.9	2.5
4	C3	VI	PHE 392D	CE1	V	3.9	2.2
4	C3	VI	THR 300D	OG1	I	4.3	1.3
6	C5	IV	PRO 272D	CD	IV	3.8	5.6
6	C5	IV	PHE 392D	CD1	V	4.0	6.5
6	C5	IV	PHE 392D	CE1	V	4.1	0.4
6	C5	IV	VAL 270D	O	II	4.3	1.6*
7	C6	IV	PRO 272D	N	VI	3.9	2.2
7	C6	IV	GLY 391D	O	II	3.9	1.6*
7	C6	IV	GLY 391D	C	VI	4.1	0.2
8	C7	IV	ALA 271D	N	III	3.3	17.5*
8	C7	IV	ALA 271D	C	VI	3.4	2.7
8	C7	IV	VAL 270D	C	VI	3.6	2.7
8	C7	IV	GLY 391D	O	II	3.9	2.0*
8	C7	IV	VAL 270D	CA	VII	4.0	2.5
8	C7	IV	VAL 270D	N	III	4.0	0.4*
8	C7	IV	PHE 392D	CA	VII	4.1	2.0
9	C8	IV	GLY 394D	N	III	3.3	28.3*
9	C8	IV	ALA 271D	O	II	3.4	18.4*
9	C8	IV	ALA 271D	C	VI	3.5	0.7
9	C8	IV	ALA 271D	N	III	3.6	3.4*
9	C8	IV	GLY 394D	CA	VI	3.6	4.9
9	C8	IV	GLY 391D	O	II	3.7	1.8*

9	C8	IV	ASP	268D	O	II	3.7	5.2*
9	C8	IV	PRO	272D	N	VI	3.8	0.4
9	C8	IV	GLY	393D	N	III	3.9	4.5*
9	C8	IV	PRO	272D	CA	IV	4.1	2.5
9	C8	IV	VAL	270D	N	III	4.1	0.2*
9	C8	IV	ASP	265D	OD2	II	4.4	0.4*
10	O1	II	HIS	333D	NE2	I	2.7	30.1
10	O1	II	HIS	298D	NE2	I	2.9	5.0
10	O1	II	CYS	163D	SG	VI	3.3	7.1
11	C9	IV	PHE	229D	CD2	V	3.5	31.9
11	C9	IV	PHE	229D	CE2	V	3.7	6.1
11	C9	IV	THR	300D	OG1	I	3.8	11.2*
11	C9	IV	THR	300D	CG2	IV	3.9	1.8
11	C9	IV	THR	302D	CG2	IV	4.0	3.6
11	C9	IV	HIS	333D	CD2	V	4.2	0.2
11	C9	IV	PHE	392D	CE1	V	4.2	7.4
11	C9	IV	PHE	392D	CZ	V	4.5	1.6
11	C9	IV	CYS	163D	SG	VI	5.1	0.9
11	C9	IV	ALA	206D	CB	IV	5.6	1.8
12	O2	I	THR	302D	OG1	I	3.9	11.6
12	O2	I	PHE	392D	CE1	V	4.0	3.8
12	O2	I	THR	302D	CG2	IV	4.4	1.0*
12	O2	I	VAL	270D	O	II	4.5	4.9
12	O2	I	THR	302D	CB	VI	4.6	0.3
12	O2	I	GLY	305D	CA	VI	4.9	0.9
12	O2	I	PRO	272D	CD	IV	5.0	1.2*
12	O2	I	GLY	305D	N	III	5.1	0.3
13	C10	IV	PRO	272D	CG	IV	3.9	16.8
13	C10	IV	GLY	305D	CA	VI	4.0	15.0
13	C10	IV	PRO	272D	CB	IV	4.4	0.4
13	C10	IV	HIS	298D	CD2	V	4.6	7.9
13	C10	IV	THR	300D	OG1	I	4.8	2.2*
13	C10	IV	PHE	390D	CB	IV	4.8	1.6
13	C10	IV	GLY	305D	C	VI	4.9	0.9
13	C10	IV	PHE	390D	CD2	V	5.0	3.6
13	C10	IV	GLY	305D	O	II	5.2	0.2*
14	C11	IV	PHE	390D	CB	IV	3.5	21.3
14	C11	IV	GLY	391D	O	II	3.5	9.0*
14	C11	IV	PRO	272D	CB	IV	3.9	10.8
14	C11	IV	PHE	390D	CG	V	4.5	0.2
14	C11	IV	ASN	396D	OD1	II	4.8	6.1*
14	C11	IV	GLY	394D	CA	VI	5.0	0.9
14	C11	IV	ASP	265D	OD2	II	5.1	0.4*

Table VI
Normalised complementarity as a function of atomic substitution for ligand TLM 503 in PDB entry 1FJ4

Legend:
N - ligand atom number in PDB entry
Bold - indicates atomic substitution which could stabilize the complex
Italics - indicates atomic substitution which could destabilize the complex

Ligand atom			Atom class							
N	Type	Class	I	II	III	IV	V	VI	VII	VIII

1	S1	VI	*0.41*	*0.31*	*0.41*	*0.29*	*0.41*	*0.41*	*0.41*	*0.31*
2	C1	VIII	*0.40*	*0.40*	*0.40*	*0.39*	*0.41*	*0.41*	*0.41*	*0.41*
3	C2	IV	*0.40*	*0.40*	*0.40*	*0.41*	*0.41*	*0.41*	*0.41*	*0.41*
4	C3	VI	*0.41*	*0.41*	*0.41*	*0.41*	*0.41*	*0.41*	*0.41*	*0.41*
6	C5	IV	*0.39*	*0.39*	*0.39*	*0.41*	*0.42*	*0.42*	*0.42*	*0.41*
7	C6	IV	*0.42*	*0.41*	*0.42*	*0.41*	*0.42*	*0.42*	*0.42*	*0.41*
8	C7	IV	0.51	0.50	*0.40*	*0.41*	0.51	0.51	*0.40*	0.50
9	C8	IV	0.71	0.58	0.53	*0.41*	0.72	0.72	0.54	0.59
10	O1	II	*0.41*	*0.41*	*0.41*	*0.24*	*0.41*	*0.41*	*0.41*	*0.41*
11	C9	IV	*0.43*	*0.43*	*0.43*	*0.41*	0.47	0.47	0.47	0.47
12	O2	I	*0.41*	*0.39*	*0.41*	*0.34*	*0.43*	*0.43*	*0.42*	*0.40*
13	C10	IV	*0.33*	*0.33*	*0.33*	*0.41*	*0.43*	*0.43*	*0.43*	*0.43*
14	C11	IV	*0.33*	*0.26*	*0.33*	*0.41*	0.49	0.49	0.49	*0.41*

Legitimacy depends on the hydrophobic-hydrophilic properties of the contacting atoms. In order to define it, for each inter-atomic contact, eight atom classes have been introduced:

I Hydrophilic - N and O that can donate and accept hydrogen bonds (e.g., oxygen of hydroxyl group of Ser. or Thr)
II Acceptor - N or O that can only accept a hydrogen bond
III Donor - N that can only donate a hydrogen bond
IV Hydrophobic - Cl, Br, I and all C atoms that are not in aromatic rings and do not have a covalent bond to a N or O atom
V Aromatic - C in aromatic rings irrespective of any other bonds formed by the atom
VI Neutral - C atoms that have a covalent bond to at least one atom of class I or two or more atoms from class II or III; atoms; S, F, P, and metal atoms in all cases
VII Neutral-donor - C atoms that have a covalent bond with only one atom of class III
VIII Neutral-acceptor - C atoms that have a covalent bond with only one atom of class II

For each pair of contacts the state of legitimacy is shown below:

Legend:
+, legitimate
-, illegitimate

Atomic class	I	II	III	IV	V	VI	VII	VIII
I (Hydrophilic)	+	+	+	-	+	+	+	+
II (Acceptor)	+	-	+	-	+	+	+	-

III (Donor)	+	+	-	-	+	+	-	+
IV (Hydrophobic)	-	-	-	+	+	+	+	+
V (Aromatic)	+	+	+	+	+	+	+	+
VI (Neutral)	+	+	+	+	+	+	+	+
VII (Neutral-donor)	+	+	-	+	+	+	-	+
VIII (Neutral-acceptor)	+	-	+	+	+	+	+	-

Contacts of ligand CER 500A in PDB entry 1FJ8

Table II
Residues in contact with ligand CER 500A in PDB entry 1FJ8
Legend:
Dist - nearest distance (Å) between atoms of the ligand and the residue
Surf - contact surface area (Å²) between the ligand and the residue
HB - hydrophilic-hydrophilic contact (hydrogen bond)
Arom - aromatic-aromatic contact
Phob - hydrophobic-hydrophobic contact
DC - hydrophobic-hydrophilic contact (destabilizing contact)
+/- - indicates presence/absence of a specific contacts
* - indicates residues contacting ligand by their side chain (including CA atoms)

Residue	Dist	Surf	HB	Arom	Phob	DC
			Specific contacts			
106A GLY*	3.2	29.2	-	-	-	+
107A GLY*	3.5	9.0	-	-	-	+
110A PRO*	4.3	9.2	-	-	+	-
162A ALA*	3.3	17.9	-	-	+	-
163A CYS*	1.9	65.1	+	-	-	+
191A GLU*	4.7	0.9	-	-	-	+
197A MET*	3.4	31.6	-	-	+	-
200A GLU*	2.8	30.3	-	-	-	+
201A PHE*	3.0	45.3	-	-	+	-
229A PHE*	5.0	7.2	-	-	+	-
298A HIS*	3.0	18.3	+	-	-	-
300A THR*	4.0	2.9	-	-	-	+
333A HIS*	2.8	25.7	+	-	+	+
335A LEU*	4.2	18.2	-	-	+	-
390A PHE*	3.7	7.4	+	-	-	+
391A GLY*	2.9	15.8	-	-	-	-
392A PHE*	2.7	72.8	+	-	+	-
113B GLN*	3.5	19.3	-	-	+	+
134B VAL*	3.3	18.1	-	-	+	+
137B ALA*	3.3	27.6	-	-	+	-
138B MET*	3.4	34.8	-	-	+	-

Table III
List of putative hydrogen bonds between ligand CER 500A and protein in PDB entry 1FJ8
Legend:
N - ligand atom number in PDB entry
Dist - distance (Å) between the ligand and protein atoms
Surf - contact surface area (Å²) between the ligand and protein atoms

Ligand atom			Protein atom			Dist	Surf
N	Name	Class	Residue	Name	Class		

N	Name	Class	Residue		Name	Class	Dist	Surf
2	N1	I	PHE	390A	O	II	3.9	5.5
2	N1	I	HIS	298A	NE2	I	4.5	1.2
5	O2	II	HIS	333A	NE2	I	2.8	21.0
5	O2	II	HIS	298A	NE2	I	3.0	17.2
7	O3	I	PHE	392A	N	III	2.7	22.3
7	O3	I	CYS	163A	N	III	2.9	7.4

Table IV
Full list of atomic contacts with ligand CER 500A in PDB entry 1FJ8
Total number of contacts is 90
Legend:
N - ligand atom number in PDB entry
Dist - distance (A) between the ligand and protein atoms
Surf - contact surface area (A**2) between the ligand and protein atoms
* - indicates destabilizing contacts

Ligand atom			Protein atom					
N	Name	Class	Residue		Name	Class	Dist	Surf
1	C1	VI	CYS	163A	SG	VI	2.8	1.8
1	C1	VI	CYS	163A	CB	IV	3.1	2.9
1	C1	VI	HIS	333A	NE2	I	3.7	0.2
1	C1	VI	PHE	390A	O	II	3.7	0.9
1	C1	VI	PHE	392A	CE1	V	4.4	0.4
1	C1	VI	HIS	333A	CD2	V	4.5	0.9
1	C1	VI	PHE	229A	CE2	V	5.2	4.0
1	C1	VI	PHE	229A	CD2	V	5.7	0.4
2	N1	I	PHE	392A	CE1	V	3.5	17.6
2	N1	I	PHE	392A	CD1	V	3.6	2.4
2	N1	I	PHE	390A	O	II	3.9	5.5
2	N1	I	GLY	391A	CA	VI	3.9	2.6
2	N1	I	GLY	391A	C	VI	4.2	0.6
2	N1	I	HIS	298A	NE2	I	4.5	1.2
2	N1	I	PHE	390A	C	VI	4.5	0.8
2	N1	I	PHE	390A	CB	IV	5.8	0.2*
3	O1	II	PHE	201A	CE2	V	3.0	20.3
3	O1	II	PHE	392A	CD2	V	3.1	14.0
3	O1	II	PHE	201A	CZ	V	3.2	0.3
3	O1	II	VAL	134B	CG1	IV	3.3	0.9*
4	C2	IV	CYS	163A	SG	VI	1.9	28.3
4	C2	IV	HIS	333A	NE2	I	3.9	1.3*
4	C2	IV	LEU	335A	CD2	IV	4.5	3.6
4	C2	IV	PHE	392A	CE1	V	4.6	0.9
4	C2	IV	HIS	333A	CD2	V	4.6	0.7
4	C2	IV	PHE	201A	CE2	V	4.7	3.4
4	C2	IV	PHE	392A	CZ	V	4.8	0.4
4	C2	IV	PHE	229A	CE2	V	5.0	2.7
5	O2	II	HIS	333A	NE2	I	2.8	21.0
5	O2	II	HIS	298A	NE2	I	3.0	17.2
5	O2	II	CYS	163A	SG	VI	3.0	0.2
5	O2	II	HIS	333A	CD2	V	3.6	1.6
5	O2	II	THR	300A	CG2	IV	4.0	2.9*
6	C3	VI	PHE	392A	N	III	3.7	5.2
6	C3	VI	PHE	392A	CD1	V	3.7	4.9

6	C3	VI	PHE	392A	CE1	V	3.9	0.4
7	O3	I	PHE	392A	N	III	2.7	22.3
7	O3	I	GLY	391A	CA	VI	2.9	12.6
7	O3	I	CYS	163A	N	III	2.9	7.4
8	C4	VIII	PHE	392A	CD2	V	3.7	4.0
8	C4	VIII	PHE	201A	CE2	V	3.8	1.6
8	C4	VIII	ALA	162A	CB	IV	3.9	1.1
9	C5	IV	CYS	163A	SG	VI	3.2	13.5
9	C5	IV	CYS	163A	N	III	3.3	11.0*
9	C5	IV	ALA	162A	CB	IV	3.3	1.6
9	C5	IV	ALA	162A	C	VI	3.4	0.9
9	C5	IV	LEU	335A	CB	IV	4.8	0.2
10	C6	IV	ALA	162A	CA	VII	3.7	12.3
10	C6	IV	ALA	162A	CB	IV	3.8	2.0
10	C6	IV	PHE	201A	CZ	V	4.0	0.2
10	C6	IV	MET	138B	CB	IV	4.1	10.5
10	C6	IV	VAL	134B	CG1	IV	4.3	0.2
10	C6	IV	MET	138B	SD	VIII	4.3	0.2
11	C7	IV	PHE	201A	CZ	V	3.4	13.5
11	C7	IV	PHE	201A	CE1	V	4.1	0.4
11	C7	IV	LEU	335A	CD1	IV	4.2	11.0
11	C7	IV	GLY	106A	C	VI	4.3	1.1
11	C7	IV	LEU	335A	CB	IV	4.4	3.4
11	C7	IV	GLY	106A	CA	VI	4.5	0.4
12	C8	IV	GLY	106A	O	II	3.2	20.4*
12	C8	IV	GLY	106A	C	VI	3.2	7.0
12	C8	IV	GLY	107A	N	III	3.5	1.6*
12	C8	IV	GLY	106A	CA	VI	3.9	0.2
12	C8	IV	MET	138B	CB	IV	4.0	0.4
12	C8	IV	MET	197A	SD	VIII	4.3	1.3
12	C8	IV	GLU	191A	OE1	II	4.7	0.9*
13	C9	IV	MET	197A	SD	VIII	3.4	20.9
13	C9	IV	GLY	107A	CA	VI	3.8	1.6
13	C9	IV	MET	197A	CG	IV	4.0	1.3
13	C9	IV	PHE	201A	CE1	V	4.0	2.9
13	C9	IV	PHE	201A	CZ	V	4.1	0.2
14	C10	IV	MET	138B	CA	VII	3.4	21.1
14	C10	IV	MET	138B	CB	IV	3.6	2.5
14	C10	IV	GLY	107A	CA	VI	4.0	5.8
14	C10	IV	VAL	134B	O	II	4.3	0.4*
14	C10	IV	VAL	134B	CG1	IV	4.5	0.4
15	C11	IV	GLU	200A	OE1	II	3.2	9.9*
15	C11	IV	MET	197A	CG	IV	4.1	7.9
15	C11	IV	VAL	134B	CA	VII	4.2	14.4
15	C11	IV	VAL	134B	CG1	IV	4.4	1.6
15	C11	IV	PHE	201A	CE1	V	4.7	2.5
15	C11	IV	VAL	134B	CG2	IV	4.7	0.2
16	C12	IV	GLU	200A	OE1	II	2.8	19.7*
16	C12	IV	ALA	137B	CB	IV	3.3	27.6
16	C12	IV	GLN	113B	NE2	III	3.5	14.6*
16	C12	IV	GLU	200A	CD	VI	3.9	0.7
16	C12	IV	GLN	113B	CG	IV	4.3	4.3
16	C12	IV	GLN	113B	CD	VI	4.3	0.4
16	C12	IV	PRO	110A	CG	IV	4.3	9.2
16	C12	IV	MET	197A	CG	IV	4.7	0.2

Table VI
Normalised complementarity as a function of atomic substitution for ligand CER 500A in PDB entry 1FJ8
Legend:

N - ligand atom number in PDB entry

Bold - indicates atomic substitution which could stabilize the complex

Italics - indicates atomic substitution which could destabilize the complex

Ligand atom			Atom class							
N	Type	Class	I	II	III	IV	V	VI	VII	VIII
1	C1	VI	0.68	0.68	0.68	0.69	0.69	0.69	0.69	0.69
2	N1	I	0.69	0.67	0.69	0.67	0.69	0.69	0.69	0.67
3	O1	II	0.69	0.69	0.69	0.70	0.70	0.70	0.70	0.70
4	C2	IV	0.68	0.68	0.68	0.69	0.70	0.70	0.70	0.70
5	O2	II	0.69	0.69	0.69	*0.55*	0.71	0.71	0.71	0.71
6	C3	VI	0.69	0.69	0.67	0.67	0.69	0.69	0.67	0.69
7	O3	I	0.69	0.69	*0.57*	*0.57*	0.69	0.69	*0.57*	0.69
8	C4	VIII	0.69	0.69	0.69	0.69	0.69	0.69	0.69	0.69
9	C5	IV	0.73	0.73	0.69	0.69	**0.74**	**0.74**	0.69	**0.74**
10	C6	IV	*0.64*	*0.64*	*0.59*	0.69	0.69	0.69	*0.64*	0.69
11	C7	IV	*0.63*	*0.63*	*0.63*	0.69	0.69	0.69	0.69	0.69
12	C8	IV	**0.79**	0.69	**0.78**	0.69	**0.79**	**0.79**	**0.78**	0.69
13	C9	IV	0.69	*0.60*	0.69	0.69	0.69	0.69	0.69	*0.61*
14	C10	IV	0.68	0.68	*0.60*	0.69	0.70	0.70	*0.61*	0.69
15	C11	IV	0.69	0.65	*0.64*	0.69	**0.73**	**0.73**	0.68	0.69
16	C12	IV	0.67	*0.58*	*0.61*	0.69	**0.83**	**0.83**	**0.77**	**0.75**

Legitimacy depends on the hydrophobic-hydrophilic properties of the contacting atoms. In order to define it, for each inter-atomic contact, eight atom classes have been introduced:

I Hydrophilic - N and O that can donate and accept hydrogen bonds (e.g., oxygen of hydroxyl group of Ser. or Thr)
II Acceptor - N or O that can only accept a hydrogen bond
III Donor - N that can only donate a hydrogen bond
IV Hydrophobic - Cl, Br, I and all C atoms that are not in aromatic rings and do not have a covalent bond to a N or O atom
V Aromatic - C in aromatic rings irrespective of any other bonds formed by the atom
VI Neutral - C atoms that have a covalent bond to at least one atom of class I or two or more atoms from class II or III; atoms; S, F, P, and metal atoms in all cases
VII Neutral-donor - C atoms that have a covalent bond with only one atom of class III
VIII Neutral-acceptor - C atoms that have a covalent bond with only one atom of class II

For each pair of contacts the state of legitimacy is shown below:

Legend:
+, legitimate
-, illegitimate

Atomic class I II III IV V VI VII VIII

I (Hydrophilic)	+	+	+	-	+	+	+	+
II (Acceptor)	+	-	+	-	+	+	+	-
III (Donor)	+	+	-	-	+	+	-	+
IV (Hydrophobic)	-	-	-	+	+	+	+	+
V (Aromatic)	+	+	+	+	+	+	+	+
VI (Neutral)	+	+	+	+	+	+	+	+
VII (Neutral-donor)	+	+	-	+	+	+	-	+
VIII (Neutral-acceptor)	+	-	+	+	+	+	+	-

Contacts of ligand CER 500B in PDB entry 1FJ8

Table II
Residues in contact with ligand CER 500B in PDB entry 1FJ8
Legend:
Dist - nearest distance (Å) between atoms of the ligand and the residue
Surf - contact surface area (Å^2) between the ligand and the residue
HB - hydrophilic-hydrophilic contact (hydrogen bond)
Arom - aromatic-aromatic contact
Phob - hydrophobic-hydrophobic contact
DC - hydrophobic-hydrophilic contact (destabilizing contact)
+/- - indicates presence/absence of a specific contacts
* - indicates residues contacting ligand by their side chain (including CA atoms)

			Specific contacts			
Residue	Dist	Surf	HB	Arom	Phob	DC
113A GLN*	3.3	27.1	-	-	+	+
133A VAL	4.6	0.7	-	-	-	+
134A VAL*	3.5	25.0	-	-	+	+
137A ALA*	3.4	16.6	-	-	+	-
138A MET*	3.6	34.3	-	-	+	-
106B GLY*	3.4	28.5	-	-	-	+
107B GLY*	3.7	7.6	-	-	-	+
110B PRO*	4.5	7.9	-	-	+	-
162B ALA*	3.3	17.3	-	-	+	-
163B CYS*	1.9	61.4	+	-	-	+
191B GLU*	4.8	0.2	-	-	-	+
197B MET*	3.4	30.1	-	-	+	+
200B GLU*	2.7	30.3	-	-	-	+
201B PHE*	2.9	44.8	-	-	+	-
229B PHE*	5.0	7.0	-	-	+	-
298B HIS*	2.7	28.8	+	-	-	-
300B THR*	4.0	2.9	-	-	-	+
333B HIS*	2.8	17.5	+	-	+	+
335B LEU*	4.2	17.3	-	-	+	-
390B PHE*	3.6	7.6	+	-	-	+
391B GLY*	2.8	16.8	-	-	-	-
392B PHE*	2.7	70.4	+	-	+	-

Table III
List of putative hydrogen bonds between ligand CER 500B and protein in PDB entry 1FJ8
Legend:
N - ligand atom number in PDB entry
Dist - distance (Å) between the ligand and protein atoms
Surf - contact surface area (Å^2) between the ligand and protein atoms

Ligand atom	Protein atom	Dist	Surf

N	Name	Class	Residue	Name	Class		
2	N1	I	PHE 390B	O	II	3.7	5.9
2	N1	I	HIS 298B	NE2	I	4.3	1.4
5	O2	II	HIS 298B	NE2	I	2.7	27.2
5	O2	II	HIS 333B	NE2	I	2.8	12.0
7	O3	I	PHE 392B	N	III	2.7	22.2
7	O3	I	CYS 163B	N	III	2.9	2.9

Table IV
Full list of atomic contacts with ligand CER 500B in PDB entry 1FJ8
Total number of contacts is 94
Legend:
N - ligand atom number in PDB entry
Dist - distance (A) between the ligand and protein atoms
Surf - contact surface area (A**2) between the ligand and protein atoms
* - indicates destabilizing contacts

Ligand atom			Protein atom				
N	Name	Class	Residue	Name	Class	Dist	Surf
1	C1	VI	CYS 163B	SG	VI	2.9	1.3
1	C1	VI	CYS 163B	CB	IV	3.0	2.9
1	C1	VI	PHE 390B	O	II	3.6	0.7
1	C1	VI	HIS 333B	NE2	I	3.7	0.7
1	C1	VI	HIS 333B	CD2	V	4.4	0.9
1	C1	VI	THR 300B	CG2	IV	5.0	0.7
1	C1	VI	PHE 229B	CE2	V	5.3	3.4
1	C1	VI	PHE 229B	CD2	V	5.8	0.2
2	N1	I	PHE 392B	CE1	V	3.6	15.8
2	N1	I	PHE 390B	O	II	3.7	5.9
2	N1	I	PHE 392B	CD1	V	3.7	2.2
2	N1	I	GLY 391B	CA	VI	3.9	3.6
2	N1	I	GLY 391B	C	VI	4.2	0.6
2	N1	I	HIS 298B	NE2	I	4.3	1.4
2	N1	I	PHE 390B	C	VI	4.4	0.6
2	N1	I	HIS 298B	CD2	V	5.2	0.2
2	N1	I	PHE 390B	CB	IV	5.6	0.4*
3	O1	II	PHE 201B	CE2	V	2.9	20.6
3	O1	II	PHE 392B	CE2	V	3.0	12.6
3	O1	II	PHE 392B	CD2	V	3.0	1.2
3	O1	II	PHE 201B	CZ	V	3.2	0.2
3	O1	II	VAL 134A	CG1	IV	3.5	0.3*
4	C2	IV	CYS 163B	SG	VI	1.9	29.6
4	C2	IV	HIS 333B	NE2	I	3.8	1.3*
4	C2	IV	HIS 333B	CD2	V	4.5	1.6
4	C2	IV	PHE 392B	CE1	V	4.6	1.1
4	C2	IV	LEU 335B	CD2	IV	4.6	2.5
4	C2	IV	PHE 392B	CZ	V	4.8	0.7
4	C2	IV	PHE 201B	CE2	V	4.8	2.2
4	C2	IV	PHE 229B	CE2	V	5.0	3.4
5	O2	II	HIS 298B	NE2	I	2.7	27.2
5	O2	II	HIS 333B	NE2	I	2.8	12.0
5	O2	II	CYS 163B	CB	IV	3.0	0.3*

5	O2	II	HIS	333B	CD2	V	3.7	1.0
5	O2	II	THR	300B	CG2	IV	4.0	2.3*
6	C3	VI	PHE	392B	N	III	3.6	4.7
6	C3	VI	PHE	392B	CD1	V	3.6	5.2
6	C3	VI	PHE	392B	CE1	V	3.8	0.4
7	O3	I	PHE	392B	N	III	2.7	22.2
7	O3	I	GLY	391B	CA	VI	2.8	12.6
7	O3	I	CYS	163B	N	III	2.9	2.9
8	C4	VIII	CYS	163B	SG	VI	3.2	0.2
8	C4	VIII	PHE	392B	CD2	V	3.7	3.8
8	C4	VIII	PHE	201B	CE2	V	3.8	2.0
8	C4	VIII	PHE	392B	CG	V	3.9	0.4
8	C4	VIII	ALA	162B	CB	IV	3.9	0.4
9	C5	IV	CYS	163B	SG	VI	3.1	13.9
9	C5	IV	CYS	163B	N	III	3.2	10.1*
9	C5	IV	ALA	162B	CB	IV	3.3	2.0
9	C5	IV	ALA	162B	C	VI	3.5	0.7
9	C5	IV	LEU	335B	CB	IV	4.8	0.2
10	C6	IV	ALA	162B	CB	IV	3.7	11.0
10	C6	IV	ALA	162B	CA	VII	3.8	3.1
10	C6	IV	PHE	201B	CZ	V	3.9	0.2
10	C6	IV	MET	138A	CB	IV	4.2	9.4
10	C6	IV	VAL	134A	CG1	IV	4.3	2.0
10	C6	IV	MET	138A	SD	VIII	4.3	0.4
11	C7	IV	PHE	201B	CZ	V	3.3	13.0
11	C7	IV	PHE	201B	CE1	V	4.0	0.4
11	C7	IV	LEU	335B	CD1	IV	4.2	10.5
11	C7	IV	LEU	335B	CB	IV	4.3	3.4
11	C7	IV	GLY	106B	C	VI	4.4	0.9
11	C7	IV	GLY	106B	CA	VI	4.6	0.9
12	C8	IV	GLY	106B	O	II	3.4	20.0*
12	C8	IV	GLY	106B	C	VI	3.4	6.5
12	C8	IV	GLY	107B	N	III	3.7	1.1*
12	C8	IV	MET	138A	CB	IV	4.0	0.9
12	C8	IV	GLY	106B	CA	VI	4.0	0.2
12	C8	IV	MET	197B	SD	VIII	4.3	1.1
12	C8	IV	LEU	335B	CD1	IV	4.8	0.7
12	C8	IV	GLU	191B	OE1	II	4.8	0.2*
13	C9	IV	MET	197B	SD	VIII	3.4	20.4
13	C9	IV	PHE	201B	CE1	V	3.8	4.7
13	C9	IV	GLY	107B	CA	VI	4.1	1.6
14	C10	IV	MET	138A	CA	VII	3.6	20.9
14	C10	IV	MET	138A	CB	IV	3.7	2.7
14	C10	IV	VAL	134A	CG1	IV	4.2	1.6
14	C10	IV	GLY	107B	CA	VI	4.2	4.9
15	C11	IV	GLU	200B	OE1	II	3.4	6.5*
15	C11	IV	VAL	134A	CA	VII	4.1	15.3
15	C11	IV	VAL	134A	CG1	IV	4.2	2.5
15	C11	IV	MET	197B	CG	IV	4.2	7.4
15	C11	IV	VAL	134A	CG2	IV	4.4	3.4
15	C11	IV	PHE	201B	CE1	V	4.4	1.3
15	C11	IV	MET	197B	O	II	4.7	0.4*
16	C12	IV	GLU	200B	OE1	II	2.7	23.3*
16	C12	IV	GLN	113A	OE1	II	3.3	25.8*
16	C12	IV	ALA	137A	CB	IV	3.4	16.6
16	C12	IV	GLU	200B	CD	VI	3.9	0.4
16	C12	IV	GLN	113A	CD	VI	4.2	1.1
16	C12	IV	GLN	113A	CG	IV	4.5	0.2

```
16  C12   IV    PRO  110B   CG    IV      4.5     7.9
16  C12   IV    VAL  133A   O     II      4.6     0.7*
16  C12   IV    MET  197B   CG    IV      4.8     0.7
-----------------------------------------------------------------------
```

Table VI
Normalised complementarity as a function of atomic substitution for ligand CER 500B in PDB entry 1FJ8

Legend:
N - ligand atom number in PDB entry
Bold - indicates atomic substitution which could stabilize the complex
Italics - indicates atomic substitution which could destabilize the complex

Ligand atom			Atom class							
N	Type	Class	I	II	III	IV	V	VI	VII	VIII
1	C1	VI	0.63	0.63	0.63	0.64	0.64	0.64	0.64	0.64
2	N1	I	0.64	0.62	0.64	0.61	0.64	0.64	0.64	0.62
3	O1	II	0.64	0.64	0.64	0.64	0.64	0.64	0.64	0.64
4	C2	IV	0.64	0.64	0.64	0.64	0.65	0.65	0.65	0.65
5	O2	II	0.64	0.64	0.64	0.49	0.65	0.65	0.65	0.65
6	C3	VI	0.64	0.64	0.62	0.62	0.64	0.64	0.62	0.64
7	O3	I	0.64	0.64	0.54	0.54	0.64	0.64	0.54	0.64
8	C4	VIII	0.64	0.64	0.64	0.64	0.64	0.64	0.64	0.64
9	C5	IV	0.67	0.67	0.63	0.64	0.68	0.68	0.64	0.68
10	C6	IV	0.55	0.55	0.54	0.64	0.64	0.64	0.63	0.64
11	C7	IV	0.59	0.59	0.59	0.64	0.64	0.64	0.64	0.64
12	C8	IV	0.72	0.64	0.72	0.64	0.73	0.73	0.72	0.64
13	C9	IV	0.64	0.56	0.64	0.64	0.64	0.64	0.64	0.56
14	C10	IV	0.63	0.63	0.54	0.64	0.64	0.64	0.56	0.64
15	C11	IV	0.62	0.59	0.55	0.64	0.67	0.67	0.61	0.64
16	C12	IV	0.74	0.54	0.74	0.64	0.85	0.85	0.85	0.64

Legitimacy depends on the hydrophobic-hydrophilic properties of the contacting atoms. In order to define it, for each inter-atomic contact, eight atom classes have been introduced:

I Hydrophilic - N and O that can donate and accept hydrogen bonds (e.g., oxygen of hydroxyl group of Ser. or Thr)
II Acceptor - N or O that can only accept a hydrogen bond
III Donor - N that can only donate a hydrogen bond
IV Hydrophobic - Cl, Br, I and all C atoms that are not in aromatic rings and do not have a covalent bond to a N or O atom
V Aromatic - C in aromatic rings irrespective of any other bonds formed by the atom
VI Neutral - C atoms that have a covalent bond to at least one atom of class I or two or more atoms from class II or III; atoms; S, F, P, and metal atoms in all cases
VII Neutral-donor - C atoms that have a covalent bond with only one

atom of class III

VIII Neutral-acceptor - C atoms that have a covalent bond with only one atom of class II

For each pair of contacts the state of legitimacy is shown below:

Legend:
+, legitimate
-, illegitimate

Atomic class	I	II	III	IV	V	VI	VII	VIII
I (Hydrophilic)	+	+	+	-	+	+	+	+
II (Acceptor)	+	-	+	-	+	+	+	-
III (Donor)	+	+	-	-	+	+	-	+
IV (Hydrophobic)	-	-	-	+	+	+	+	+
V (Aromatic)	+	+	+	+	+	+	+	+
VI (Neutral)	+	+	+	+	+	+	+	+
VII (Neutral-donor)	+	+	-	+	+	+	-	+
VIII (Neutral-acceptor)	+	-	+	+	+	+	+	-

Contacts of ligand CER 500C in PDB entry 1FJ8

Table II
Residues in contact with ligand CER 500C in PDB entry 1FJ8
Legend:
Dist - nearest distance (Å) between atoms of the ligand and the residue
Surf - contact surface area (Å^2) between the ligand and the residue
HB - hydrophilic-hydrophilic contact (hydrogen bond)
Arom - aromatic-aromatic contact
Phob - hydrophobic-hydrophobic contact
DC - hydrophobic-hydrophilic contact (destabilizing contact)
+/- - indicates presence/absence of a specific contacts
* - indicates residues contacting ligand by their side chain
(including CA atoms)

			Specific contacts			
Residue	Dist	Surf	HB	Arom	Phob	DC
106C GLY*	3.1	31.9	-	-	-	+
107C GLY*	3.6	7.2	-	-	-	+
110C PRO*	4.3	7.6	-	-	+	-
162C ALA*	3.3	15.9	-	-	+	-
163C CYS*	1.9	58.4	-	-	-	+
191C GLU*	4.9	0.2	-	-	-	+
197C MET*	3.3	33.2	-	-	+	-
200C GLU*	2.9	28.9	-	-	-	+
201C PHE*	2.7	49.8	-	-	+	-
229C PHE*	4.8	7.7	-	-	+	-
298C HIS*	2.8	20.4	+	-	-	-
300C THR*	3.8	3.8	+	-	-	+
333C HIS*	2.6	29.4	+	-	+	+
335C LEU*	4.3	17.9	-	-	+	-
390C PHE*	3.8	5.9	+	-	-	+
391C GLY*	2.9	10.1	-	-	-	-
392C PHE*	2.7	67.6	+	-	+	-
113D GLN*	2.9	34.5	-	-	+	+
134D VAL*	3.4	19.2	-	-	+	+
137D ALA*	3.3	11.7	-	-	+	-
138D MET*	3.4	35.7	-	-	+	+

Table III
List of putative hydrogen bonds between ligand CER 500C and protein in PDB entry 1FJ8
Legend:
N - ligand atom number in PDB entry
Dist - distance (Å) between the ligand and protein atoms
Surf - contact surface area (Å^2) between the ligand and protein atoms

Ligand atom			Protein atom			Dist	Surf
N	Name	Class	Residue	Name	Class		

N	Name	Class	Residue	Name	Class	Dist	Surf
2	N1	I	PHE 390C	O	II	4.0	4.0
2	N1	I	HIS 298C	NE2	I	4.3	1.8
2	N1	I	THR 300C	OG1	I	5.8	0.2
5	O2	II	HIS 333C	NE2	I	2.6	21.1
5	O2	II	HIS 298C	NE2	I	2.8	18.0
7	O3	I	PHE 392C	N	III	2.7	24.9

Table IV
Full list of atomic contacts with ligand CER 500C in PDB entry 1FJ8
Total number of contacts is 92
Legend:
N - ligand atom number in PDB entry
Dist - distance (A) between the ligand and protein atoms
Surf - contact surface area (A**2) between the ligand and protein atoms
* - indicates destabilizing contacts

Ligand atom			Protein atom			Dist	Surf
N	Name	Class	Residue	Name	Class		
1	C1	VI	CYS 163C	SG	VI	2.8	1.3
1	C1	VI	CYS 163C	CB	IV	3.1	2.5
1	C1	VI	HIS 333C	NE2	I	3.4	1.1
1	C1	VI	PHE 390C	O	II	3.8	1.3
1	C1	VI	HIS 333C	CD2	V	4.2	2.0
1	C1	VI	THR 300C	CG2	IV	4.9	0.9
1	C1	VI	PHE 229C	CE2	V	5.0	3.4
1	C1	VI	PHE 229C	CD2	V	5.5	0.4
1	C1	VI	THR 300C	OG1	I	5.7	0.2
2	N1	I	PHE 392C	CE1	V	3.6	15.6
2	N1	I	PHE 392C	CD1	V	3.9	2.0
2	N1	I	PHE 390C	O	II	4.0	4.0
2	N1	I	HIS 298C	NE2	I	4.3	1.8
2	N1	I	GLY 391C	CA	VI	4.3	2.0
2	N1	I	PHE 390C	C	VI	4.8	0.4
2	N1	I	HIS 298C	CD2	V	5.2	0.6
2	N1	I	THR 300C	CG2	IV	5.4	0.4*
2	N1	I	PHE 229C	CE2	V	5.4	1.0
2	N1	I	THR 300C	OG1	I	5.8	0.2
2	N1	I	PHE 390C	CB	IV	6.0	0.2*
3	O1	II	PHE 201C	CE2	V	2.7	27.9
3	O1	II	PHE 201C	CZ	V	2.9	0.3
3	O1	II	PHE 392C	CE2	V	3.1	9.7
3	O1	II	PHE 392C	CD2	V	3.2	0.3
3	O1	II	VAL 134D	CG1	IV	3.4	0.2*
4	C2	IV	CYS 163C	SG	VI	1.9	28.9
4	C2	IV	HIS 333C	NE2	I	3.6	2.2*
4	C2	IV	HIS 333C	CD2	V	4.4	1.8
4	C2	IV	LEU 335C	CD2	IV	4.4	3.1
4	C2	IV	PHE 392C	CE1	V	4.7	0.2
4	C2	IV	PHE 201C	CE2	V	4.7	2.5
4	C2	IV	PHE 229C	CE2	V	4.8	2.9
5	O2	II	HIS 333C	NE2	I	2.6	21.1
5	O2	II	HIS 298C	NE2	I	2.8	18.0
5	O2	II	CYS 163C	SG	VI	3.0	0.7
5	O2	II	HIS 333C	CD2	V	3.4	1.0
5	O2	II	THR 300C	CG2	IV	3.8	2.1*

6	C3	VI	PHE	392C	CD1	V	3.7	8.7
6	C3	VI	PHE	392C	CE1	V	3.8	1.3
6	C3	VI	PHE	392C	CZ	V	4.0	0.2
7	O3	I	PHE	392C	N	III	2.7	24.9
7	O3	I	GLY	391C	CA	VI	2.9	8.1
7	O3	I	CYS	163C	CB	IV	3.1	0.5*
8	C4	VIII	CYS	163C	SG	VI	3.2	0.4
8	C4	VIII	PHE	201C	CE2	V	3.6	1.3
8	C4	VIII	PHE	392C	CD2	V	3.7	4.5
9	C5	IV	CYS	163C	SG	VI	3.2	12.8
9	C5	IV	CYS	163C	N	III	3.2	11.2*
9	C5	IV	ALA	162C	CB	IV	3.3	2.5
9	C5	IV	ALA	162C	C	VI	3.5	0.4
10	C6	IV	ALA	162C	CB	IV	3.7	11.7
10	C6	IV	ALA	162C	CA	VII	3.8	1.3
10	C6	IV	PHE	201C	CZ	V	3.9	0.4
10	C6	IV	MET	138D	CB	IV	4.1	9.4
10	C6	IV	MET	138D	SD	VIII	4.2	2.2
10	C6	IV	VAL	134D	CG1	IV	4.2	1.1
11	C7	IV	PHE	201C	CZ	V	3.5	11.7
11	C7	IV	PHE	201C	CE1	V	4.2	0.4
11	C7	IV	GLY	106C	C	VI	4.2	1.8
11	C7	IV	LEU	335C	CD1	IV	4.3	10.8
11	C7	IV	LEU	335C	CB	IV	4.4	4.0
11	C7	IV	GLY	106C	CA	VI	4.4	1.3
12	C8	IV	GLY	106C	O	II	3.1	22.7*
12	C8	IV	GLY	106C	C	VI	3.2	5.8
12	C8	IV	GLY	107C	N	III	3.6	0.7*
12	C8	IV	MET	138D	CB	IV	3.8	1.3
12	C8	IV	GLY	106C	CA	VI	3.9	0.2
12	C8	IV	MET	197C	SD	VIII	4.4	0.2
12	C8	IV	GLU	191C	OE1	II	4.9	0.2*
13	C9	IV	MET	197C	SD	VIII	3.3	21.5
13	C9	IV	GLY	107C	CA	VI	3.8	2.0
13	C9	IV	MET	197C	CG	IV	4.0	0.7
13	C9	IV	PHE	201C	CE1	V	4.1	2.5
14	C10	IV	MET	138D	CA	VII	3.4	20.0
14	C10	IV	MET	138D	CB	IV	3.6	2.5
14	C10	IV	MET	138D	N	III	3.7	0.2*
14	C10	IV	GLY	107C	CA	VI	4.1	4.5
14	C10	IV	VAL	134D	O	II	4.2	1.1*
14	C10	IV	VAL	134D	CG1	IV	4.3	1.3
15	C11	IV	GLU	200C	OE1	II	3.4	7.4*
15	C11	IV	MET	197C	CG	IV	3.8	10.5
15	C11	IV	VAL	134D	CA	VII	4.3	13.0
15	C11	IV	VAL	134D	CG1	IV	4.4	2.0
15	C11	IV	PHE	201C	CE1	V	4.4	2.7
15	C11	IV	VAL	134D	CG2	IV	4.7	0.4
16	C12	IV	GLN	113D	NE2	III	2.9	32.5*
16	C12	IV	GLU	200C	OE1	II	2.9	21.5*
16	C12	IV	ALA	137D	CB	IV	3.3	11.7
16	C12	IV	GLN	113D	CD	VI	3.9	1.6
16	C12	IV	GLN	113D	CG	IV	4.2	0.4
16	C12	IV	PRO	110C	CG	IV	4.3	7.6
16	C12	IV	MET	197C	CG	IV	4.3	0.2

Table VI
Normalised complementarity as a function of atomic substitution for ligand CER 500C in PDB entry 1FJ8

Legend:

N - ligand atom number in PDB entry

Bold - indicates atomic substitution which could stabilize the complex

Italics - indicates atomic substitution which could destabilize the complex

Ligand atom			Atom class							
N	Type	Class	I	II	III	IV	V	VI	VII	VIII
1	C1	VI	0.59	0.58	0.59	0.59	0.60	0.60	0.60	0.59
2	N1	I	0.60	0.58	0.60	0.58	0.60	0.60	0.60	0.59
3	O1	II	0.60	0.60	0.60	0.60	0.60	0.60	0.60	0.60
4	C2	IV	0.60	0.60	0.60	0.60	0.61	0.61	0.61	0.61
5	O2	II	0.60	0.60	0.60	0.45	0.61	0.61	0.61	0.61
6	C3	VI	0.60	0.60	0.60	0.60	0.60	0.60	0.60	0.60
7	O3	I	0.60	0.60	0.50	0.50	0.60	0.60	0.50	0.60
8	C4	VIII	0.60	0.60	0.60	0.60	0.60	0.60	0.60	0.60
9	C5	IV	0.64	0.64	0.59	0.60	0.65	0.65	0.60	0.65
10	C6	IV	0.51	0.50	0.50	0.60	0.60	0.60	0.59	0.59
11	C7	IV	0.54	0.54	0.54	0.60	0.60	0.60	0.60	0.60
12	C8	IV	0.69	0.60	0.69	0.60	0.70	0.70	0.69	0.60
13	C9	IV	0.60	0.51	0.60	0.60	0.60	0.60	0.60	0.51
14	C10	IV	0.59	0.58	0.51	0.60	0.61	0.61	0.52	0.60
15	C11	IV	0.58	0.55	0.52	0.60	0.63	0.63	0.58	0.60
16	C12	IV	0.74	0.65	0.61	0.60	0.82	0.82	0.69	0.73

Legitimacy depends on the hydrophobic-hydrophilic properties of the contacting atoms. In order to define it, for each inter-atomic contact, eight atom classes have been introduced:

I Hydrophilic - N and O that can donate and accept hydrogen bonds (e.g., oxygen of hydroxyl group of Ser. or Thr)

II Acceptor - N or O that can only accept a hydrogen bond

III Donor - N that can only donate a hydrogen bond

IV Hydrophobic - Cl, Br, I and all C atoms that are not in aromatic rings and do not have a covalent bond to a N or O atom

V Aromatic - C in aromatic rings irrespective of any other bonds formed by the atom

VI Neutral - C atoms that have a covalent bond to at least one atom of class I or two or more atoms from class II or III; atoms; S, F, P, and metal atoms in all cases

VII Neutral-donor - C atoms that have a covalent bond with only one atom of class III

VIII Neutral-acceptor - C atoms that have a covalent bond with only one atom of class II

For each pair of contacts the state of legitimacy is shown below:

Legend:
+, legitimate
-, illegitimate

Atomic class	I	II	III	IV	V	VI	VII	VIII
I (Hydrophilic)	+	+	+	-	+	+	+	+
II (Acceptor)	+	-	+	-	+	+	+	-
III (Donor)	+	+	-	-	+	+	-	+
IV (Hydrophobic)	-	-	-	+	+	+	+	+
V (Aromatic)	+	+	+	+	+	+	+	+
VI (Neutral)	+	+	+	+	+	+	+	+
VII (Neutral-donor)	+	+	-	+	+	+	-	+
VIII (Neutral-acceptor)	+	-	+	+	+	+	+	-

Contacts of ligand CER 500D in PDB entry 1FJ8

Table II
Residues in contact with ligand CER 500D in PDB entry 1FJ8
Legend:
Dist - nearest distance (Å) between atoms of the ligand and the residue
Surf - contact surface area ($Å^2$) between the ligand and the residue
HB - hydrophilic-hydrophilic contact (hydrogen bond)
Arom - aromatic-aromatic contact
Phob - hydrophobic-hydrophobic contact
DC - hydrophobic-hydrophilic contact (destabilizing contact)
+/- - indicates presence/absence of a specific contacts
* - indicates residues contacting ligand by their side chain
(including CA atoms)

Residue	Dist	Surf	HB	Arom	Phob	DC
			Specific contacts			
113C GLN*	4.3	11.0	-	-	+	-
134C VAL*	3.5	15.9	-	-	+	+
137C ALA*	3.4	24.7	-	-	+	+
138C MET*	3.6	32.1	-	-	+	-
106D GLY*	3.3	30.1	-	-	-	+
107D GLY*	3.6	7.6	-	-	-	+
110D PRO*	4.4	7.6	-	-	+	-
162D ALA*	3.3	16.6	-	-	+	-
163D CYS*	1.9	58.6	-	-	-	+
191D GLU*	4.8	0.7	-	-	-	+
197D MET*	3.4	38.8	-	-	+	+
200D GLU*	2.8	34.5	-	-	-	+
201D PHE*	2.8	49.9	-	-	+	-
229D PHE*	4.9	6.3	-	-	+	-
298D HIS*	2.8	20.2	+	-	-	-
300D THR*	3.9	3.4	-	-	-	+
333D HIS*	2.7	28.5	+	-	+	+
335D LEU*	4.2	17.9	-	-	+	-
390D PHE*	3.7	6.7	+	-	-	+
391D GLY*	2.8	15.1	-	-	-	-
392D PHE*	2.7	70.3	+	-	+	-

Table III
List of putative hydrogen bonds between ligand CER 500D and protein in PDB entry 1FJ8
Legend:
N - ligand atom number in PDB entry
Dist - distance (Å) between the ligand and protein atoms
Surf - contact surface area ($Å^2$) between the ligand and protein atoms

Ligand atom			Protein atom			Dist	Surf
N	Name	Class	Residue	Name	Class		

2	N1	I	PHE 390D	O	II	3.9	4.8
2	N1	I	HIS 298D	NE2	I	4.3	1.8
5	O2	II	HIS 333D	NE2	I	2.7	21.1
5	O2	II	HIS 298D	NE2	I	2.8	18.2
7	O3	I	PHE 392D	N	III	2.7	24.1

Table IV
Full list of atomic contacts with ligand CER 500D in PDB entry 1FJ8
Total number of contacts is 92
Legend:
N - ligand atom number in PDB entry
Dist - distance (A) between the ligand and protein atoms
Surf - contact surface area (A**2) between the ligand and protein atoms
* - indicates destabilizing contacts

Ligand atom			Protein atom				
N	Name	Class	Residue	Name	Class	Dist	Surf
1	C1	VI	CYS 163D	SG	VI	2.9	1.8
1	C1	VI	CYS 163D	CB	IV	3.1	2.2
1	C1	VI	HIS 333D	NE2	I	3.5	0.9
1	C1	VI	PHE 390D	O	II	3.7	1.3
1	C1	VI	HIS 333D	CD2	V	4.3	1.6
1	C1	VI	THR 300D	CG2	IV	4.9	0.9
1	C1	VI	PHE 229D	CE2	V	5.1	3.6
1	C1	VI	THR 300D	OG1	I	5.8	0.2
2	N1	I	PHE 392D	CE1	V	3.7	16.4
2	N1	I	PHE 392D	CD1	V	3.8	2.0
2	N1	I	PHE 390D	O	II	3.9	4.8
2	N1	I	GLY 391D	CA	VI	4.1	2.6
2	N1	I	HIS 298D	NE2	I	4.3	1.8
2	N1	I	GLY 391D	C	VI	4.4	0.2
2	N1	I	PHE 390D	C	VI	4.6	0.4
2	N1	I	HIS 298D	CD2	V	5.2	0.2
2	N1	I	PHE 390D	CB	IV	5.8	0.2*
3	O1	II	PHE 201D	CE2	V	2.8	23.2
3	O1	II	PHE 392D	CE2	V	3.1	11.3
3	O1	II	PHE 392D	CD2	V	3.1	0.3
3	O1	II	VAL 134C	CG1	IV	3.5	0.2*
4	C2	IV	CYS 163D	SG	VI	1.9	28.7
4	C2	IV	HIS 333D	NE2	I	3.7	2.0*
4	C2	IV	HIS 333D	CD2	V	4.4	1.8
4	C2	IV	LEU 335D	CD2	IV	4.5	2.9
4	C2	IV	PHE 392D	CE1	V	4.7	0.4
4	C2	IV	PHE 201D	CE2	V	4.8	2.7
4	C2	IV	PHE 392D	CZ	V	4.9	0.7
4	C2	IV	PHE 229D	CE2	V	4.9	2.7
5	O2	II	HIS 333D	NE2	I	2.7	21.1
5	O2	II	HIS 298D	NE2	I	2.8	18.2
5	O2	II	CYS 163D	CB	IV	3.1	0.7*
5	O2	II	HIS 333D	CD2	V	3.5	1.0
5	O2	II	THR 300D	CG2	IV	3.9	2.3*
6	C3	VI	PHE 392D	CD1	V	3.7	9.6
6	C3	VI	PHE 392D	CE1	V	3.8	0.9

7	O3	I	PHE	392D	N	III	2.7	24.1
7	O3	I	GLY	391D	CA	VI	2.8	12.3
7	O3	I	CYS	163D	CB	IV	3.1	0.7*
8	C4	VIII	CYS	163D	SG	VI	3.2	0.2
8	C4	VIII	PHE	392D	CD2	V	3.7	4.0
8	C4	VIII	PHE	201D	CE2	V	3.7	1.6
8	C4	VIII	PHE	392D	CG	V	4.0	0.4
9	C5	IV	CYS	163D	SG	VI	3.1	13.5
9	C5	IV	CYS	163D	N	III	3.2	10.8*
9	C5	IV	ALA	162D	CB	IV	3.3	2.0
9	C5	IV	ALA	162D	C	VI	3.5	0.7
10	C6	IV	ALA	162D	CB	IV	3.7	11.2
10	C6	IV	ALA	162D	CA	VII	3.7	2.7
10	C6	IV	PHE	201D	CZ	V	3.9	0.2
10	C6	IV	MET	138C	CB	IV	4.2	8.7
10	C6	IV	VAL	134C	CG1	IV	4.3	2.2
10	C6	IV	MET	138C	SD	VIII	4.3	0.4
11	C7	IV	PHE	201D	CZ	V	3.4	11.9
11	C7	IV	PHE	201D	CE1	V	4.1	0.2
11	C7	IV	LEU	335D	CD1	IV	4.2	11.0
11	C7	IV	LEU	335D	CB	IV	4.3	4.0
11	C7	IV	GLY	106D	C	VI	4.3	1.3
11	C7	IV	GLY	106D	CA	VI	4.5	1.1
12	C8	IV	GLY	106D	O	II	3.3	21.5*
12	C8	IV	GLY	106D	C	VI	3.3	6.1
12	C8	IV	GLY	107D	N	III	3.6	0.9*
12	C8	IV	MET	138C	CB	IV	3.9	0.9
12	C8	IV	MET	197D	SD	VIII	4.4	0.7
12	C8	IV	GLU	191D	OE1	II	4.8	0.7*
13	C9	IV	MET	197D	SD	VIII	3.4	20.6
13	C9	IV	PHE	201D	CE1	V	3.9	4.3
13	C9	IV	GLY	107D	CA	VI	4.0	1.1
14	C10	IV	MET	138C	CA	VII	3.6	19.5
14	C10	IV	MET	138C	CB	IV	3.7	2.5
14	C10	IV	GLY	107D	CA	VI	4.2	5.2
14	C10	IV	VAL	134C	O	II	4.3	1.1*
14	C10	IV	VAL	134C	CG1	IV	4.3	1.3
15	C11	IV	GLU	200D	OE1	II	3.4	7.9*
15	C11	IV	MET	197D	CG	IV	3.8	11.2
15	C11	IV	VAL	134C	CA	VII	4.3	10.3
15	C11	IV	GLU	200D	CD	VI	4.4	0.2
15	C11	IV	PHE	201D	CE1	V	4.4	5.8
15	C11	IV	VAL	134C	CG2	IV	4.6	0.7
15	C11	IV	MET	197D	O	II	4.7	0.2*
16	C12	IV	GLU	200D	OE1	II	2.8	21.3*
16	C12	IV	ALA	137C	CB	IV	3.4	22.7
16	C12	IV	GLU	200D	CD	VI	3.8	4.0
16	C12	IV	GLU	200D	OE2	II	4.0	1.1*
16	C12	IV	ALA	137C	C	VI	4.1	1.1
16	C12	IV	MET	197D	CG	IV	4.3	5.8
16	C12	IV	ALA	137C	O	II	4.3	0.9*
16	C12	IV	GLN	113C	CG	IV	4.3	10.5
16	C12	IV	PRO	110D	CG	IV	4.4	7.6
16	C12	IV	MET	197D	CB	IV	4.8	0.2
16	C12	IV	GLN	113C	CD	VI	5.0	0.4
16	C12	IV	GLY	107D	CA	VI	5.4	0.4

Table VI
Normalised complementarity as a function of atomic substitution for ligand CER 500D in PDB entry 1FJ8
Legend:

N - ligand atom number in PDB entry

Bold - indicates atomic substitution which could stabilize the complex

Italics - indicates atomic substitution which could destabilize the complex

Ligand atom			Atom class							
N	Type	Class	I	II	III	IV	V	VI	VII	VIII
1	C1	VI	0.71	0.70	0.71	0.71	0.72	0.72	0.72	0.72
2	N1	I	0.72	0.70	0.72	0.70	0.72	0.72	0.72	0.70
3	O1	II	0.72	0.72	0.72	0.72	0.72	0.72	0.72	0.72
4	C2	IV	0.72	0.72	0.72	0.72	0.73	0.73	0.73	0.73
5	O2	II	0.72	0.72	0.72	0.57	0.73	0.73	0.73	0.73
6	C3	VI	0.72	0.72	0.72	0.72	0.72	0.72	0.72	0.72
7	O3	I	0.72	0.72	0.62	0.63	0.73	0.73	0.63	0.73
8	C4	VIII	0.72	0.72	0.72	0.72	0.72	0.72	0.72	0.72
9	C5	IV	0.76	0.76	0.71	0.72	0.77	0.77	0.72	0.77
10	C6	IV	0.63	0.63	0.62	0.72	0.72	0.72	0.71	0.72
11	C7	IV	0.66	0.66	0.66	0.72	0.72	0.72	0.72	0.72
12	C8	IV	0.81	0.72	0.81	0.72	0.82	0.82	0.81	0.72
13	C9	IV	0.72	0.64	0.72	0.72	0.72	0.72	0.72	0.64
14	C10	IV	0.71	0.71	0.63	0.72	0.73	0.73	0.65	0.72
15	C11	IV	0.71	0.67	0.66	0.72	0.76	0.76	0.71	0.72
16	C12	IV	0.63	0.53	0.63	0.72	0.82	0.82	0.82	0.72

Legitimacy depends on the hydrophobic-hydrophilic properties of the contacting atoms. In order to define it, for each inter-atomic contact, eight atom classes have been introduced:

I Hydrophilic - N and O that can donate and accept hydrogen bonds (e.g., oxygen of hydroxyl group of Ser. or Thr)
II Acceptor - N or O that can only accept a hydrogen bond
III Donor - N that can only donate a hydrogen bond
IV Hydrophobic - Cl, Br, I and all C atoms that are not in aromatic rings and do not have a covalent bond to a N or O atom
V Aromatic - C in aromatic rings irrespective of any other bonds formed by the atom
VI Neutral - C atoms that have a covalent bond to at least one atom of class I or two or more atoms from class II or III; atoms; S, F, P, and metal atoms in all cases
VII Neutral-donor - C atoms that have a covalent bond with only one atom of class III
VIII Neutral-acceptor - C atoms that have a covalent bond with only one atom of class II

For each pair of contacts the state of legitimacy is shown below:

Legend:
+, legitimate
-, illegitimate

Atomic class	I	II	III	IV	V	VI	VII	VIII
I (Hydrophilic)	+	+	+	-	+	+	+	+
II (Acceptor)	+	-	+	-	+	+	+	-
III (Donor)	+	+	-	-	+	+	-	+
IV (Hydrophobic)	-	-	-	+	+	+	+	+
V (Aromatic)	+	+	+	+	+	+	+	+
VI (Neutral)	+	+	+	+	+	+	+	+
VII (Neutral-donor)	+	+	-	+	+	+	-	+
VIII (Neutral-acceptor)	+	-	+	+	+	+	+	-

Contacts of ligand PO4 340 in PDB entry 1HNJ

Table II
Residues in contact with ligand PO4 340 in PDB entry 1HNJ
Legend:
Dist - nearest distance (Å) between atoms of the ligand and the residue
Surf - contact surface area (Å[2]) between the ligand and the residue
HB - hydrophilic-hydrophilic contact (hydrogen bond)
Arom - aromatic-aromatic contact
Phob - hydrophobic-hydrophobic contact
DC - hydrophobic-hydrophilic contact (destabilizing contact)
+/- - indicates presence/absence of a specific contacts
* - indicates residues contacting ligand by their side chain
(including CA atoms)

Residue	Dist	Surf	HB	Arom	Phob	DC
			Specific contacts			
16A VAL*	2.8	29.3	+	-	-	+
18A THR*	4.0	15.4	+	-	-	+
21A ASP*	2.8	32.4	+	-	-	-

Table III
List of putative hydrogen bonds between ligand PO4 340 and protein in PDB entry 1HNJ
Legend:
N - ligand atom number in PDB entry
Dist - distance (Å) between the ligand and protein atoms
Surf - contact surface area (Å[2]) between the ligand and protein atoms

Ligand atom			Protein atom				Dist	Surf
N	Name	Class	Residue		Name	Class		
3	O2	I	VAL	16A	O	II	4.7	0.2
4	O3	I	VAL	16A	O	II	2.8	27.5
4	O3	I	ASP	21A	OD1	II	2.8	12.6
5	O4	I	ASP	21A	OD2	II	3.0	19.4
5	O4	I	THR	18A	OG1	I	4.6	2.8

Table IV
Full list of atomic contacts with ligand PO4 340 in PDB entry 1HNJ
Total number of contacts is 11
Legend:
N - ligand atom number in PDB entry
Dist - distance (A) between the ligand and protein atoms
Surf - contact surface area (A**2) between the ligand and protein atoms
* - indicates destabilizing contacts

Ligand atom			Protein atom			Dist	Surf
N	Name	Class	Residue	Name	Class		

```
---------------------------------------------------------------------
 2  O1   I    THR  18A   CG2  IV     4.6    6.8*
 2  O1   I    VAL  16A   CB   IV     5.9    0.5*
 3  O2   I    VAL  16A   O    II     4.7    0.2
 4  O3   I    VAL  16A   O    II     2.8   27.5
 4  O3   I    ASP  21A   OD1  II     2.8   12.6
 4  O3   I    THR  18A   CG2  IV     4.0    5.7*
 4  O3   I    VAL  16A   CB   IV     4.9    1.0*
 5  O4   I    ASP  21A   OD2  II     3.0   19.4
 5  O4   I    ASP  21A   CG   VI     3.7    0.3
 5  O4   I    THR  18A   OG1  I      4.6    2.8
 5  O4   I    THR  18A   CG2  IV     5.0    0.2*
---------------------------------------------------------------------
```

Table VI
Normalised complementarity as a function of atomic substitution for ligand PO4 340 in PDB entry 1HNJ

Legend:
N - ligand atom number in PDB entry
Bold - indicates atomic substitution which could stabilize the complex
Italics - indicates atomic substitution which could destabilize the complex

Ligand atom			Atom class							
N	Type	Class	I	II	III	IV	V	VI	VII	VIII
2	O1	I	0.25	0.25	0.25	0.32	0.32	0.32	0.32	0.32
3	O2	I	0.25	0.24	0.25	0.24	0.25	0.25	0.25	0.24
4	O3	I	0.25	*-0.16*	0.25	*-0.09*	0.31	0.31	0.31	*-0.09*
5	O4	I	0.25	*0.05*	0.25	*0.02*	0.25	0.25	0.25	*0.05*

Legitimacy depends on the hydrophobic-hydrophilic properties of the contacting atoms. In order to define it, for each inter-atomic contact, eight atom classes have been introduced:

I Hydrophilic - N and O that can donate and accept hydrogen bonds (e.g., oxygen of hydroxyl group of Ser. or Thr)
II Acceptor - N or O that can only accept a hydrogen bond
III Donor - N that can only donate a hydrogen bond
IV Hydrophobic - Cl, Br, I and all C atoms that are not in aromatic rings and do not have a covalent bond to a N or O atom
V Aromatic - C in aromatic rings irrespective of any other bonds formed by the atom
VI Neutral - C atoms that have a covalent bond to at least one atom of class I or two or more atoms from class II or III; atoms; S, F, P, and metal atoms in all cases
VII Neutral-donor - C atoms that have a covalent bond with only one atom of class III
VIII Neutral-acceptor - C atoms that have a covalent bond with only one atom of class II
For each pair of contacts the state of legitimacy is shown below:

Legend:
+, legitimate
-, illegitimate

Atomic class	I	II	III	IV	V	VI	VII	VIII
I (Hydrophilic)	+	+	+	-	+	+	+	+
II (Acceptor)	+	-	+	-	+	+	+	-
III (Donor)	+	+	-	-	+	+	-	+
IV (Hydrophobic)	-	-	-	+	+	+	+	+
V (Aromatic)	+	+	+	+	+	+	+	+
VI (Neutral)	+	+	+	+	+	+	+	+
VII (Neutral-donor)	+	+	-	+	+	+	-	+
VIII (Neutral-acceptor)	+	-	+	+	+	+	+	-

2.8 BIBLIOGRAPHY

1.Verh K Acad Geneeskd Belg. (1999). Infections with flaviviridae. 61(6):661-97; discussion 697-9.

2.María G. Guzmán and Gustavo Kourí (2004). Dengue diagnosis, advances and challenges. International Journal of Infectious Diseases, Volume 8, Issue 2, Pages 69-80

3.Charles H. Calisher and Ernest A. Gould (2003). Taxonomy of the virus family Flaviviridae Advances in Virus Research, Volume 59, Pages 1-19

4.Kadare, G. & Haenni, A.L. (1997). Virus-encoded RNA helicases. J. Virol. 71, 2583–2590.

5.Francoise Degos (1994). Epidemiology of hepatitis C virus in Europe. FEMS Microbiology Reviews, Volume 14, Issue 3, Pages 267-271

6.L. Pasta, C. Marrone, M. D'amico, E.A. Bevacqua, G. D'Amico and L. Pagliaro (2005) Familial clustering of HCV-related liver cirrhosis and hepatocellular carcinoma. Digestive and Liver Disease, Volume 37, Issue 9, Pages 716-717

7.Marie-Pierre Courageot , Adeline Catteau, and Philippe Desprès (2003). Mechanisms of Dengue virus-induced cell death. Advances in Virus Research, Volume 60, Pages 157-186

8.Stephen J. Thomas , Daniel Strickman, and David W. Vaughn (2003). Dengue Epidemiology: Virus Epidemiology, Ecology, and Emergence. Advances in Virus Research, Volume 61 , Pages 235-289

9.Sadaf Shadan and BMN News (2004). Unfolding targets for dengue fever. Drug Discovery Today, Volume 9, Issue 8, Page 344

10. Tian Wang and Erol Fikrig (2004). Immunity to West Nile virus. Current Opinion in Immunology, Volume 16, Issue 4, Pages 519-523

11. Marc S. Collett (1992). Molecular genetics of pestiviruses. Comparative Immunology, Microbiology and Infectious Diseases, Volume 15, Issue 3, Pages 145-154

12. Ramiro Avalos-Ramirez, Michaela Orlich, Heinz-Jürgen Thiel and Paul Becher (2001), Evidence for the Presence of Two Novel Pestivirus Species. Virology, Volume 286, Issue 2, 1 August 2001, Pages 456-465
13. Colin W Shepard, Lyn Finelli and Miriam J Alter (2005). Global epidemiology of hepatitis C virus infection. The Lancet Infectious Diseases, Volume 5, Issue 9, Pages 558-567
14. K. Venugopal and E. A. Gould (1994). Towards a new generation of flavivirus vaccines. Vaccine, Volume 12, Issue 11, Pages 966-975
15. Walker, M. 1999. Hepatitis C virus: an overview of current approaches and progress. Drug Discovery Today. Vol.4, No.11
16. Sudo, K. Matsumoto, Y. Matsushima, M. Fujiwara, M. Konno, K. Shimotohno, K. Shigeta, S. and Yokota, T. 1997. Novel Hepatitis C Virus Protease Inhibitors: Thiazolidine Derivatives. Biochemical and Biophysical Research Communications (238)
17. Hepatitis C virus ultrastructure and morphogenesis. Biology of the Cell, Volume 96, Issue 2, March 2004, Pages 103-108
18. Kim, J.L., et al., & Thomson, J.A. (1996). Crystal structure of the hepatitis C virus NS3 protease domain complexed with a synthetic NS4A cofactor peptide. Cell 87, 343–355.
19. Yao, N., Hesson, T., Cable, M., Hong, Z., Kwong, A.D., Le, H.V. & Weber, P.C. (1997). Structure of the hepatitis C virus RNA helicace domain. Nat. Struct. Biol. 4, 463–467.
20. Jean-Pierre Allain (2005). Hepatitis C virus in blood donation. The Lancet, Volume 365, Issue 945, Pages 276-278
21. Lisa K. Gilbert, Jane Bulger, Kelli Scanlon and Linda Moyer (2005) Viral hepatitis prevention education: what do people and providers need to know? Patient Education and Counseling, Volume 59, Issue 1, Pages 46-55

22. Emmet B. Keeffe (2005). Chronic Hepatitis C: Management of Treatment Failures. Clinical Gastroenterology and Hepatology, Volume 3, Supplement 2, Pages S102-S105

23. Marion Riffelmann, Silke Sarr, Antje Ballauff, Helga Meisel and Michael Roggendorf (1997). Transmission of GBV-C/HGV from drug-addicted mothers to their babies. Journal of Hepatology, Volume 27, Issue 1, July 1997, Pages 85-90

24. Beerntsen B. T., Severson D. W., Klinkhammer J. A., Kassner V. A. and Christensen B. M. (1995). Aedes aegypti: A Quantitative Trait Locus (QTL) Influencing Filarial Worm Intensity Is Linked to QTL for Susceptibility to Other Mosquito-Borne Pathogens. Experimental Parasitology, Volume 81, Issue 3, Pages 355-362

25. Daryl T.-Y. Lau and Alex T. Hewlett (2005) Screening for hepatitis A and B antibodies in patients with chronic liver disease. The American Journal of Medicine, Volume 118, Issue 10, Supplement 1, October 2005, Pages 28-33

26. Suzich, J.A., et al., & Collett, M.S. (1993). Hepatitis C virus NS3 protein polynucleotide-stimulated nucleoside triphosphatase and comparison with the related pestivirus and flavivirus enzymes. J. Virol. 67, 6152–6158.

27. Tsukiyama, K. K., N. Iizuka, M. Kohara, and A. Nomoto. 1992. Internal ribosome entry site within hepatitis C virus RNA. J. Virol. 66:1476–1483.

28. Pickering, J. M., H. C. Thomas, and P. Karayiannis. 1997. Predicted secondary structure of the hepatitis G virus and GB virus-A 5$untranslated regions consistent with an internal ribosome entry site. J. Viral Hepat. 4:175–184

29. Miriam J. Alter (2006) Epidemiology of viral hepatitis and HIV co-infection Journal of Hepatology, Volume 44, Supplement 1, Pages S6-S9

30. Chen, Y., T. Maguire, R. E. Hileman, J. R. Fromm, J. D. Esko, R. J. Linhardt, and R. M. Marks. 1997. Dengue virus infectivity depends on envelope protein binding to target cell heparan sulfate. Nat. Med. 3:866–871

31. Pryor, M. J., R. C. Gualano, B. Lin, A. D. Davidson, and P. J. Wright. 1998. Growth restriction of dengue virus type 2 by site-specific mutagenesis of virus-encoded glycoproteins. J. Gen. Virol. 79:2631–2639.
32. Wengler, G., and G. Wengler. 1993. The NS3 nonstructural protein of flaviviruses contains an RNA triphosphatase activity. Virology 197:265–273.
33. Styer LM, Carey JR, Wang JL, Scott TW. (2007) Mosquitoes do senesce: departure from the paradigm of constant mortality. Am J Trop Med Hyg. 76(1):111-7.
34. Khan AM, Heiny A, Lee KX, Srinivasan K, Tan TW, August JT, Brusic V. (2006) Large-scale analysis of antigenic diversity of T-cell epitopes in dengue virus. BMC Bioinformatics. 18;7 Suppl 5:S4.
35. Chareonsirisuthigul T, Kalayanarooj S, Ubol S. (2007) Dengue virus (DENV) antibody-dependent enhancement of infection upregulates the production of anti-inflammatory cytokines, but suppresses anti-DENV free radical and pro-inflammatory cytokine production, in THP-1 cells. J Gen Virol.88(Pt 2):365-75.
36. Suksanpaisan L, Cabrera-Hernandez A, Smith DR. (2007) Infection of human primary hepatocytes with dengue virus serotype 2. J Med Virol. 23;79(3):300-307
37. Ang KT, Ruhaini I, Chua KB. (2006) An epidemiological cluster pattern of dengue outbreak amongst close contacts in Selangor, Peninsular Malaysia. Med J Malaysia. 61(3):292-5.
38. Thomas P Monath (2001). Yellow fever: an update. The Lancet Infectious Diseases, Volume 1, Issue 1, Pages 11-20
39. Stefan Schilling, Diana Ludolfs, Le Van An and Herbert Schmitz (2004). Laboratory diagnosis of primary and secondary dengue infection. Journal of Clinical Virology, Volume 31, Issue 3, Pages 179-184
40. Edwards, S., P. M. Roehe, and G. Ibata. 1995. Comparative studies of border disease and closely related virus infections in experimental pigs and sheep. Br. Vet. J. 151:181–187.

41. Terpstra, C., and G. Wensvoort. 1997. A congenital persistent infection of bovine virus diarrhoea virus in pigs: clinical, virological and immunological observations. Vet. Q. 19:97–101

42. Van Campen, H., E. S. Williams, J. Edwards, W. Cook, and G. Stout. 1997. Experimental infection of deer with bovine viral diarrhea virus. J. Wildl. Dis. 33:567–573.

43. Meehan, J. T., H. D. Lehmkuhl, R. C. Cutlip, and S. R. Bolin. 1998. Acute pulmonary lesions in sheep experimentally infected with bovine viral diarrhoea virus. J. Comp. Pathol. 119:277–292.

44. Pastoret, P. P., and M. Lambot. 1995. Biologie de l'infection des bovins par le Pestivirus responsable de la maladie des muqueuses. Bull. Mem. Acad. R. Med. Belg. 150:244–250. (In French.)

45. Thiel, H.-J., P. G. W. Plagemann, and V. Moennig. 1996. Pestiviruses, p. 1059–1073. In B. N. Fields, D. M. Knipe, and P. M. Howley (ed.), Fields virology, 3rd ed., vol. 1. Lippincott-Raven Publishers, Philadelphia, Pa.

46. Bruschke, C. J., K. Weerdmeester, J. T. Van Oirschot, and P. A. Van Rijn. 1998. Distribution of bovine virus diarrhoea virus in tissues and white blood cells of cattle during acute infection. Vet. Microbiol. 64:23–32.

47. Ellis, J. A., K. H. West, V. S. Cortese, S. L. Myers, S. Carman, K. M. Martin, and D. M. Haines. 1998. Lesions and distribution of viral antigen following an experimental infection of young seronegative calves with virulent bovine virus diarrhea virus-type II. Can. J. Vet. Res. 62:161–169.

48. Donis, R. O., and E. J. Dubovi. 1987. Differences in virus-induced polypeptides in cells infected by cytopathic and noncytopathic biotypes of bovine virus diarrhea-mucosal disease virus. Virology 158:168–173.

49. Kummerer, B. M., D. Stoll, and G. Meyers. 1998. Bovine viral diarrhea virus strain Oregon: a novel mechanism for processing of NS2-3 based on point mutations. J. Virol. 72:4127–4138.

50. Mendez, E., N. Ruggli, M. S. Collett, and C. M. Rice. 1998. Infectious bovine viral diarrhea virus (strain NADL) RNA from stable cDNA clones: a cellular insert determines NS3 production and viral cytopathogenicity. J. Virol. 72:4737–4745.
51. Meyers, G., and H. J. Thiel. 1996. Molecular characterization of pestiviruses. Adv. Virus Res. 47:53–118
52. Tautz, N., G. Meyers, and H.-J. Thiel. 1998. Pathogenesis of mucosal disease, a deadly disease of cattle caused by a pestivirus. Clin. Diagn. Virol. 10:121–127.
53. Makoto Nagai, Michiko Hayashi, Shigeo Sugita, Yoshihiro Sakoda, Masashi Mori, Toshiaki Murakami, Tadashi Ozawa, Naoki Yamada and Hiroomi Akashi (2004). Phylogenetic analysis of bovine viral diarrhea viruses using five different genetic regions. Virus Research, Volume 99, Issue 2, February 2004, Pages 103-113
54. Laevens, H., H. Deluyker, F. Koenen, G. Van Caenegem, J. P. Vermeersch, and A. de Kruif. 1998. An experimental infection with a classical swine fever virus in weaner pigs. II. The use of serological data to estimate the day of virus introduction in natural outbreaks. Vet. Q. 20:46–49
55. Kim, J.L., et al., & Thomson, J.A. (1996). Crystal structure of the hepatitis C virus NS3 protease domain complexed with a synthetic NS4A cofactor peptide. Cell 87, 343–355.
56. Laevens, H., F. Koenen, H. Deluyker, D. Berkvens, and A. de Kruif. 1998. An experimental infection with a classical swine fever virus in weaner pigs. I. Transmission of the virus, course of the disease, and antibody response. Vet. Q. 20:41–45.
57. Vilcek, S., T. Stadejek, P. A. Ballagi, J. P. Lowings, D. J. Paton, and S. Belak. 1996. Genetic variability of classical swine fever virus. Virus Res. 43:137–147.
58. Nettleton, P. F., J. A. Gilray, P. Russo, and E. Dlissi. 1998. Border disease of sheep and goats. Vet. Res. 29:327–340.
59. Becher, P., G. Meyers, A. D. Shannon, and H. J. Thiel. 1996. Cytopathogenicityof border disease virus is correlated with integration of cellular sequences into the viral genome. J. Virol. 70:2992–2998.

60. Bolin, S. R. 1995. Control of bovine viral diarrhea infection by use of vaccination. Vet. Clin. North Am. Food Anim. Pract. 11:615–625.

61. Cortese, V. S., D. L. Grooms, J. Ellis, S. R. Bolin, J. F. Ridpath, and K. V. Brock. 1998. Protection of pregnant cattle and their fetuses against infection with bovine viral diarrhea virus type 1 by use of a modified-live virus vaccine. Am. J. Vet. Res. 59:1409–1413.

62. Cortese, V. S., R. Whittaker, J. Ellis, J. F. Ridpath, and S. R. Bolin. 1998. Specificity and duration of neutralizing antibodies induced in healthy cattle after administration of a modified-live virus vaccine against bovine viral diarrhea. Am. J. Vet. Res. 59:848–850.

63. Poole, T. L., C. Wang, R. A. Popp, L. N. Potgieter, A. Siddiqui, and M. S. Collett. 1995. Pestivirus translation initiation occurs by internal ribosome entry. Virology 206:750–754.

64. R. L. Doherty (1958). Effects of yellow fever (17D) and West Nile viruses on the reactions of human appendix and conjunctiva cells to several other viruses. Virology, Volume 6, Issue 3, Pages 575-583

65. David N. Phalen and Bob Dahlhausen (2004). West Nile virus. Seminars in Avian and Exotic Pet Medicine, Volume 13, Issue 2, Pages 67-78

66. Blaithnead Murtagh, Yasmin Wadia, Greg Messner, Paul Allison, Yadollah Harati and Reynolds Delgado (2005). West Nile Virus Infection After Cardiac Transplantation. The Journal of Heart and Lung Transplantation, Volume 24, Issue 6, June 2005, Pages 774-776

67. Thompson, B. and Finch, R. 2005. Hepatitis C Virus Infection. Clinical Microbiology and Infection 11(2), pp. 86-94.

68. Rosenberg, S. 2001. Recent Advances in the Molecular Biology of Hepatitis C Virus. Journal of Molecular Biology 313, pp. 451-464.

69. Poynard, T. Bedossa, P. Opolon, P.1997. Natural history of liver fibrosis progression in patients with chronic hepatitis C. Lancet 349, pp. 825-832.

70. Daly, M. and Ward, F. 2003. Liver Disease. In: Walker, R. Edwards, C. (Ed) Clinical Pharmacy and Therapeutics 3rd ed. Churchill Livingstone. pp. 209-228.
71. Pontisso, P. Belluco, C. Bertorelle, R. Moliner, L. Bianchi, L. Nitti, D. Lise, M. Alberti, A. 1998. Hepatitis C Virus Infection Associated with Human Hepatocellular Carcinoma. Cancer 83(8), pp. 1489-1494.
72. Triyatni, M. Vergalla, J. Davis, A. Hadlock, K. Foung, S. and Liang, T. 2002. Structural Features of Envelope Proteins on Hepatitis C Virus-like Particles as Determined by Anti-envelope Monoclonal Antibodies and CD81 Binding. Virology 298, pp. 124-132.
73. Keller, P. Gordon, C. 2005. Control of Hepatitis C: A Medicinal Chemistry Perspective. Journal of Medicinal Chemistry 48(1), pp 1-20
74. Steffens, S. Thiel, H. and Behrens, S.1999. The RNA-dependent RNA polymerases of different members of the family Flaviviridae exhibit similar properties in vitro. Journal of General Virology 80, pp. 2583-2590.
75. Heidemarie Holzmann (2003). Diagnosis of tick-borne encephalitis. Vaccine, Volume 21, Supplement 1, Pages S36-S40
76. Lewis Markoff (2000). Points to consider in the development of a surrogate for efficacy of novel Japanese encephalitis virus vaccines. Vaccine, Volume 18, Supplement 2, Pages 26-32
77. Denise S. Olive, Masayoshi Konishi and George Y. Wu (2004). Cell culture and animal models for human viral hepatitis. Hepatology Research, Volume 28, Issue 2, Pages 61-67
78. Caruthers, J. & McKay, D. Helicase structure and mechanism. Curr. Opin. Struct. Biol. 12, 123–133 (2002).
79. Pascal Pineau , Anne Dejean and Pierre Tiollais (2002). Animal hepadnaviruses and their host species: Models for human hepatocarcinogenesis. Perspectives in Medical Virology, Volume 6, Pages 123-141

80. Gwack, Y., Kim, D.W., Han, J.H. & Choe, J. (1996). Characterization of RNA binding activity and RNA helicase activity of the hepatitis C virus NS3 protein. Biochem. Biophys. Res. Commun. 225, 654–659.

81. Behrens, S. E., L. Tomei, and R. De Francesco. 1996. Identification and properties of the RNA-dependent RNA polymerase of hepatitis C virus. EMBO J. 15:12–22

82. Wengler, G. and Wengler, G. 1993. The NS3 Nonstructural Protein of Flaviviruses Contains an RNA Triphosphatase Activity. Virology 197

83. Leyssen, P. Clercq, E. and Neyts, J. 2000. Perspectives for the Treatment of Infections with Flaviviridae. Clinical Microbiology Reviews 13(1), pp. 67-82.

84. Tan, S. He, Y. Huang, Y. and Gale, M. 2004. Strategies for hepatitis C therapeutic intervention: now and next. Current Opinion in Pharmacology 4, pp. 465-470.

85. Walker, M. 1999. Hepatitis C virus: an overview of current approaches and progress. Drug Discovery Today. Vol.4, No.11, pp 518-529

86. Nulf, C. J. and Corey, D. 2004. Intracellular inhibition of hepatitis C virus (HCV)internal ribosomal entry site (IRES)-dependent translation by peptide nucleic acids (PNAs) and locked nucleic acids (LNAs). Nucleic Acids Research 32(13), pp. 3792-3798.

87. Jefferson,E. Seth, P. Robinson, D. Winter, D. Miyaji, A. Osgood, S. Swayze, E. and Risen, L.2004. Biaryl guanidine inhibitors of in vitro HCV-IRES activity. Bioorganic Medicinal Chemistry Letters 14, pp. 5139-5143.

88. Wengler, G. and Wengler, G. 1993. The NS3 Nonstructural Protein of Flaviviruses Contains an RNA Triphosphatase Activity. Virology 197, pp. 265-273.

89. Sudo, K. Matsumoto, Y. Matsushima, M. Fujiwara, M. Konno, K. Shimotohno, K.Shigeta, S. and Yokota, T. 1997. Novel Hepatitis C Virus Protease Inhibitors:Thiazolidine Derivatives. Biochemical and Biophysical Research Communications (238) pp 643–647

90. Ingallinella, P. Altamura, S. Bianchi, E. Taliani, M. Ingenito, R. Cortese, R. De Francesco, R. Steinkühler, C. and Antonello Pessi 1998. Potent Peptide Inhibitors of Human Hepatitis C Virus NS3 Protease Are Obtained by Optimizing the Cleavage Products. Biochemistry 37(25) pp 8906 - 8914.
91. Benarroch, D. Egloff, M. Mulard, L. Guerreiro, C. Romette, J. and Canard, B. 2004. A Structural Basis for the Inhibition of the NS5 Dengue Virus mRNA 2'-OMethyltransferase Domain by Ribavirin 5'-Triphosphate. The Journal of Biological Chemistry 279(34), pp. 35638-35643.
92. Chang, S. Cheng, J. Kou,Y. Kao,C. Chiu, C. Wu, H. and Chang, M. 2000. Roles of the AX4GKS and Arginine-Rich Motifs of Hepatitis C Virus RNA Helicase in ATP- and Viral RNA-Binding Activity. Journal of Virology 74(20), pp. 9732-9737.
93. Artsaenko, O. Tessmann, K. Sack, M. Haussinger, D. Heintges, T. 2003. Abrogation of hepatitis C virus NS3 helicase enzymatic activity by recombinant human antibodies. Journal of General Virology 84, pp 2323–2332.
94. Matusan, A. Pryor, M. Davidson, A. Wright, P. 2001. Mutagenesis of the Dengue VirusType 2 NS3 Protein within and outside Helicase Motifs: Effects on Enzyme Activity and Virus Replication. Journal of Virology, pp 9633–9643
95. Lam, A. Rypma, R. Frick, D. 2004. Enhanced nucleic acid binding to ATP-bound hepatitis C virus NS3 helicase at low pH activates RNA unwinding. Nucleic AcidsResearch, Vol. 32, No. 13, pp 4060-4070
96. Bianco, P. 2004. Hepatitis C NS3 helicase unwinds RNA in leaps and bounds. The Lancet. Vol. 364, pp 1385-1387 Borowski, P. Schalinski, S. Schmitz, H. 2002. Nucleotide triphosphatase/helicase of hepatitis C virus as a target for antiviral therapy. Antiviral Research (55) pp 397-412
97. Phoon, C. Ng, P. Ting, A. Yeo, S. Sim, M. 2001. Biological Evaluation of Hepatitis C Virus Helicase Inhibitors. Bioorganic & Medicinal Chemistry Letters (11) pp 1647–1650
98. Diana, G. Bailey, T. 1997. Compounds, compositions and methods for treatment ofhepatitis C. United States Patent No. 5,633,388

99. ViroFarma Patent: Compounds, Compositions and Methods for treatment of Hepatitis C, Patent number: 5.633.388, Date of Patent: May 27, 1997

100. Steffens, S. Thiel, H. and Behrens, S.1999. The RNA-dependent RNA polymerases of different members of the family Flaviviridae exhibit similar properties in vitro. Journal of General Virology 80

101. Chang, S. Cheng, J. Kou,Y. Kao,C. Chiu, C. Wu, H. and Chang, M. (2000). Roles of the AX4GKS and Arginine-Rich Motifs of Hepatitis C Virus RNA Helicase in ATP- and Viral RNA-Binding Activity. Journal of Virology 74(20), 9732-9737.

102. Lam, A. Rypma, R. Frick, D. (2004). Enhanced nucleic acid binding to ATP-bound hepatitis C virus NS3 helicase at low pH activates RNA unwinding. Nucleic Acids Research, Vol. 32, No. 13, 4060-4070

103. Artsaenko, O. Tessmann, K. Sack, M. Haussinger, D. Heintges, T. (2003). Abrogation of hepatitis C virus NS3 helicase enzymatic activity by recombinant human antibodies. Journal of General Virology 84, 2323–2332.

104. Matusan, A. Pryor, M. Davidson, A. Wright, P. (2001). Mutagenesis of the Dengue Virus Type 2 NS3 Protein within and outside Helicase Motifs: Effects on Enzyme Activity and Virus Replication. Journal of Virology, pp 9633–9643

105. Jin, L. & Peterson, D.L. (1995). Expression, isolation, and characterization of the hepatitis C virus ATPase/RNA helicase. Arch. Biochem. Biophys. 323, 47–53.

106. Preugschat, F., Averett, D.R., Clarke, B.E. & Porter, D.J.T. (1996). A steady-state and pre-steady-state kinetic analysis of the NTPase activity associated with the hepatitis C virus NS3 helicase domain. J. Biol. Chem. 271, 24449–24457.

107. Morgenstern, K.A., et al., & Thomson, J.A. (1997). Polynucleotide modulation of the protease, nucleoside triphosphatase, and helicase activities of a hepatitis C virus NS3–NS4A complex isolated from transfected COS cells. J. Virol. 71, 3767–3775.

108. Hong, Z., et al., & Kwong, A.D. (1996). Enzymatic characterization of hepatitis C virus NS3/4A complexes expressed in mammalian cells by using the herpes simplex virus amplicon system. J. Virol. 70,4261–4268.

109. M.P. Allen and D.J. Tildesley. Computer Simulation of Liquids. Oxford University Press, New York, 1987.

110. A.D. Bates and A. Maxwell. DNA Topology. In Focus series, Oxford Univeristy Press, New York, 1993.

111. C. Branden and J. Tooze. Introduction to Protein Structure. second edition, Garland Publishing Inc., New York, 1999.

112. P. Bratley, B. L. Fox, and L. E. Schrage. A Guide to Simulation. Springer-Verlag, New York, 1987.

113. C. L. Brooks, III, M. Karplus, and B.M. Pettitt. A Theoretical Perspective of Dynamics, Structure, and Thermodynamics. Wiley Interscience, New York, 1988.

114. U. Burkert and N.L. Allinger. Molecular Mechanics. American Chemical Society, Washington D.C., 1980.

115. C.R. Cantor and P.R. Schimmel. Biophysical Chemistry. Vol.1,2,3. W.H. Freeman and Company, San Fransisco, 1980.

116. N. R. Cohen, Editor. Guidebook on Molecular Modeling in Drug Design. Academic Press, San Diego, 1996.

117. Computational Molecular Dynamics: Challenges, Methods, Ideas -- Proceedings of the 2nd International Symposium on Algorithms for Macromolecular Modelling, Berlin, May 21-24, 1997, P. Deuflhard, J. Hermans, B. Leimkuhler, A. E. Mark, S. Reich,. R. D. Skeel, Eds., Lecture Notes in Computational Science and Engineering, Vol. 4 (Series Editors M. Griebel, D. E. Keyes, R. M. Nieminen, D. Roose, and T. Schlick), Springer-Verlag, Berlin and New York, 1999.

118. T.E. Creighton, Editor. Protein Folding. W.H. Freeman & Company, New York, 1992.

119. D. Eisenberg and D. Crothers. Physcial Chemistry with Applications to the Life Science. Benjamin Cummings, Menlo Park, California, 1979.

120. A. Fersht. Structure and Mechanism in Protein Science: A Guide to Enzyme Catalysis and Protein Folding. W. H. Freeman and Company, New York, 1999.

121. D. Frenkel and B. Smit. Understanding Molecular Simulations. From Algorithms to Applications. Academic Press, San Diego, California, 1996.

122. L.M. Gierasch and J. King, Editors. Protein Folding, Deciphering the Second Half of the Genetic Code. AAAS, Washington D.C., 1990.

123. H. Gould and J. Tobochnik. An Introduction to Computer Simulation Metheods: Appkications to Physical Systems. Part 1 and 2 Addison-Wesley, Reading, MA, 1988.

124. A.Y. Grosberg and A.R. Khokhlov. Giant Molecules. Here, There, and Everywhere... Academic Press, San Diego, California, 1997.

125. J.M. Haile. Molecular Dynamics Simulations: Elementary Methods. Wiley, New York, 1992.

126. M. Kalos and P. A. Whitlock. Monte Carlo Methods. John Wiley & Sons, New York, 1986.

127. A. R. Leach. Molecular Modelling. Principles and Applications. Addison Wesley Longman, Essex, England, 1996.

128. K.B. Lipkowitz and D.B. Boyd, Editors. Reviews in Computational Chemistry. VCH Publishers, New York, 1990 –

129. J.A. McCammon and S.C. Harvey. Dynamics of Proteins and Nucleic Acids. Cambridge University Press, Cambridge, 1987.

130. D.A. McQuarrie. Statistical Mechanics. Harper Collins Publishers, New York, 1976.

131. National Research Council report. Mathematical Challenges from Theoretical / Computational Chemistry, National Academy Press, Washington D.C., 1995.

132. D.C. Rapaport. The Art of Molecular Dynamics Simulation. Cambridge University Press, Cambridge, England., 1995.

133. W. Saenger. Principles of Nucleic Acid Structure. Springer Advanced Texts in Chemistry, Springer-Verlag, New York, 1984.

134. R.R. Sinden. DNA Structure and Function. cademic Press, San Diego, California, 1994

135. G.E. Schulz and R.H. Schirmer. Principles of Protein Structure. Springer Advanced Texts in Chemistry, Springer-Verlag, New York, 1990.

136. L. Stryer. Biochemistry. W.H. Freeman, New York, latest edition (fourth in 1995).

137. W. Van Gunsteren and P. Weiner, Editors (1989) and W. Van Gunsteren, P. Weiner, and A.T. Wilkinson, Editors (1993, 1996): Computer Simulation of Biomolecular Systems: Theoretical and Experimentational Applications. Vol. 1,2,3. ESCOM, Leiden, The Netherlands, 1989, 1993, 1996.

138. Ding, J., Das, K., Moereels, H., Koymans, L., Andries, K., Janssen, P.A.J., Hughes, S.H., Arnold, E. Nat. Struct. Biol. 2, 407, (1995).

139. Eriksson, M.A.L., Pitera, J., Kollman, P.A. J. Med. Chem. 42, 868-881, (1999).

140. Potter, M.J., Kirchhoff, P.D., Carlson, H.A., McCammon, J.A. J. Comput. Chem. 20, 956-970, (1999).

141. Gill, P., Murray W., Wright, M.Practical Optimization. Academic Press, London, 1981.

142. Stewart, J. J. P. MOPAC Manual (Seventh Edition), 1993.

143. Allen, M.P., Tildesley, D.J. Computer Simulation of Liquids. Oxford University Press (1987).

144. Verlet, L. Computer 'Experiments' on Classical Fluids. I. Thermodynamical Properties of Lennard-Jones Molecules. Phys. Rev. 159, 98-103 (1967).

145. Bond, S.D., Benedict, J.L., Laird, B.B. The Nosé-Poincaré Method for Constant Temperature Molecular Dynamics. J. Comp. Phys. 151, 114-134 (1999).

146. Sturgeon, J.B., Laird, B.B. Symplectic Algorithm for Constant Pressure Molecular Dynamics Using a Nosé-Poincaré Thermostat. University of Kansas Technical Paper (2002).

147. Sun, G. Symplectic Partitioned Runge-Kutta Methods. J. Comput. Math. 11, 365-372 (1993).

148. Berendsen, H.J.C., Postma, J.P.M., Van Gunsteren, W.F., Di Nola, A., Haak, J.R. Molecular Dynamics with Coupling to an External Bath. J. Chem. Phys. 81, 3684-3690 (1984).

149. Ryckaert, J.P., Ciccotti, G., Berendsen, H.J.C. Numerical Integration of the Cartesian Equations of Motion of a System with Constraints: Molecular Dynamics of n-Alkanes. J. Comput. Phys. 23, 327-341 (1977).

150. Hess, B., Bekker, H., Berendsen, H.J.C., Fraaije, J.G.E.M. LINCS: A Linear Constraint Solver for Molecular Simulations J. Comp. Chem. 18, 1463-1472 (1997).

151. Ferguson, D.M., Raber, D.J. A New Approach to Probing Conformational Space with Molecular Mechanics: Random Incremental Pulse Search. J. Am. Chem. Soc. 111, 4371-4378 (1989).

152. H.M.Berman, J.Westbrook, Z.Feng, G.Gilliland, T.N.Bhat, H.Weissig, I.N.Shindyalov, P.E.Bourne. The Protein Data Bank. Nucleic Acids Research. 28, 235-242, (2000).

153. Fechteler T., Dengler, U., Schomberg, D. Prediction of Protein Three-Dimensional Structures in Insertion and Deletion Regions: A Procedure for Searching Databases of Representative Protein Fragments Using Geometric Scoring Criteria. Journal of Molecular Biology. 253, 114-131, (1995).

154. Jones T.A. & Thirup, S. Using Known Substructures in Protein Modeling and Crystallography. EMBO J.. 5, 819-822, (1986).

155. LPC: Ligand-protein Contacts, 2003, URL: http://www.weizmann.ac.il/sgedg/lpc/

156. LigPlot: Schematic diagrams of protein-ligand interactions, 2002, URL: http://www.biochem.ucl.ac.uk/bsm/ligplot/ligplot.html

157. Raghunathan, S., Ricard, C.S., Lohman, T.M. & Waksman, G. (1997). Crystal structure of the homo-tetrameric DNA-binding domain of Escherichia coli single-stranded DNA-binding protein determined by multiwavelength X-ray diffraction on the selenomethionyl protein at 2.9 Å resolution. Proc. Natl. Acad. Sci. USA 94, 6652–6657.

158. Gwack, Y., Kim, D.W., Han, J.H. & Choe, J. (1996). Characterization of RNA binding activity and RNA helicase activity of the hepatitis C virus NS3 protein. Biochem. Biophys. Res. Commun. 225, 654–659.

159. Kanai, A., Tanabe, K. & Kohara, M. (1995). Poly(U) binding activity of hepatitis C virus NS3 protein, a putative RNA helicase. FEBS Lett. 376, 221–224.

160. Miller, R.H. & Purcell, R.H. (1990). Hepatitis C virus shares amino acid sequence similarity with pestiviruses and flaviviruses as well as members of two plant virus supergroups. Proc. Natl. Acad. Sci. USA 87, 2057–2061.

161. Sybyl 7.0 Tripos Inc, 1699 South Hanley Road, St. Louis, Missouri, 63144, USA, URL: http://www.tripos.com

162. Mirzayan, C. & Wimmer, E. (1992). Genetic analysis of an NTPbinding motif in poliovirus polypeptide 2C. Virology 189, 547–555.

163. MacPherson, P., Thorner, L., Parker, L.M. & Botchan, M. (1994). The bovine papilloma virus E1 protein has ATPase activity essential to viral DNA replication and efficient transformation in cells. Virology 204, 403–408.

164. MOE, Molecular Operating Environment 2005.04, Chemical Computing Group, Montreal, Quebec, Canada, URL:http://www.chemcomp.com/

165. Procheck: Roman A Laskowski, Malcolm W MacArthur, David S Moss and Janet M Thornton, J. App. Cryst. 26 283 (1993)

166. Martinez, R., Shao, L. & Weller, S.K. (1992). The conserved helicase motifs of the herpes simplex virus type 1 origin-binding protein UL9 are important for function. J. Virol. 66, 6735–6746.

167. Gorbalenya, A. & Koonin, E.V. (1993). Helicases: amino acid sequence comparisons and structure-function relationships. Curr. Opin. Struct. Biol. 3, 419–429.

168. Subramanya, H.S., Bird, L.E., Brannigan, J.A. & Wigley, D.B. (1996). Crystal structure of a DExx box DNA helicase. Nature 384, 379–383.

169. Kim, J.L., Morgenstern, K.A., Griffith, J.P., Dwyer, M.D., Thomson, J.A., Murcko, M.A., Lin, C., Caron, P.R. Hepatitis C virus NS3 RNA helicase domain with a bound oligonucleotide: the crystal structure provides insights into the mode of unwinding. Structure v6 pp.89-100 , 1998

170. Horscroft, N. Lai, V. Cheney, W. Yao, N. Wu, JZ. Hong, Z. Zhong, W. 2005. Replicon cell culture system as a valuable tool in antiviral drug discovery against hepatitis C virus. Antiviral Chemistry & Chemotherapy; 16: 1–12

171. Korolev, S., Hsieh, J., Gauss, G.H., Lohman, T.M. & Waksman, G. (1997). Major domain swiveling revealed by the crystal structures of complexes of E. coli Rep helicase bound to single-stranded DNA and ADP. Cell 90, 635–647.

172. Bartenschlager, R., Ahlborn-Laake, L., Mous, J. & Jacobsen, H. (1994). Kinetic and structural analyses of hepatitis C virus polyprotein processing. J. Virol. 68, 5045–5055.

173. Lohman, T.M. & Bjornson, K.P. (1996). Mechanisms of helicasecatalyzed DNA unwinding. Annu. Rev. Biochem. 65, 169–214.

174. Walker, J.E., Saraste, M., Runswick, M.J. & Gay, N.J. (1982). Distantly related sequences in the alpha- and beta-subunits of ATP synthase, myosin, kinases and other ATP-requiring enzymes and a common nucleotide binding fold. EMBO J. 1, 945–951.

175. Saraste, M., Sibbald, P.R. & Wittinghofer, A. (1990). The P-loop – a common motif in ATP- and GTP-binding proteins. Trends Biochem. Sci. 15, 430–434.

176. Ruff, M., et al., & Moras, D. (1991). Class II aminoacyl transfer RNA synthetases: crystal structure of yeast aspartyl-tRNA synthetase complexed with tRNA(Asp). Science 252, 1682–1689.

177. Bochkarev, A., Pfuetzner, R.A., Edwards, A.M. & Frappier, L. (1997). Structure of the single-stranded-DNA-binding domain of replication protein A bound to DNA. Nature 385, 176–181.

178. Kim, Y., Geiger, J.H., Hahn, S. & Sigler, P.B. (1993). Crystal structure of a yeast TBP/TATA-box complex. Nature 365, 512–520.

179. Kim, J.L., Nikolov, D.B. & Burley, S.K. (1993). Co-crystal structure of TBP recognizing the minor groove of a TATA element. Nature 365, 520–527.

180. George, J.W., Brosh, R.M., Jr. & Matson, S.W. (1994). A dominant negative allele of the Escherichia coli uvrD gene encoding DNA helicase II. A biochemical and genetic characterization. J. Mol. Biol.

181. Seeley, T.W. & Grossman, L. (1990). The role of Escherichia coli UvrB in nucleotide excision repair. J. Biol. Chem. 265, 7158–7165.

182. Black, M.E. & Hruby, D.E. (1992). Site-directed mutagenesis of a conserved domain in vaccinia virus thymidine kinase. Evidence for a potential role in magnesium binding. J. Biol. Chem. 267, 6801–6806.

183. Brosh, R.M., Jr. & Matson, S.W. (1995). Mutations in motif II of Escherichia coli DNA helicase II render the enzyme nonfunctional in both mismatch repair and excision repair with differential effects on the unwinding reaction. J. Bacteriol. 177, 5612–5621.

184. Pause, A. & Sonenberg, N. (1992). Mutational analysis of a DEAD box RNA helicase: the mammalian translation initiation factor eIF-4A. EMBO J. 11, 2643–2654.

185. Heilek, G.M. & Peterson, M.G. (1997). A point mutation abolishes the helicase but not the nucleoside triphosphatase activity of hepatitis C virus NS3 protein. J. Virol. 71, 6264–6266.

186. Gross, C.H. & Shuman, S. (1996). The QRxGRxGRxxxG motif of the vaccinia virus DExH box RNA helicase NPH- II is required for ATP hydrolysis and RNA unwinding but not for RNA binding. J. Virol. 70, 1706–1713.

187. Pause, A., Methot, N. & Sonenberg, N. (1993). The HRIGRXXR region of the DEAD box RNA helicase eukaryotic translation initiation factor 4A is required for RNA binding and ATP hydrolysis. Mol. Cell Biol. 13, 6789–6798.

188. Hall, M.C. & Matson, S.W. (1997). Mutation of a highly conserved arginine in motif IV of Escherichia coli DNA helicase II results in an ATP-binding defect. J. Biol. Chem. 272, 18614–18620.

189. Bilderback, T., Fulmer, T., Mantulin, W.W. & Glaser, M. (1996). Substrate binding causes movement in the ATP binding domain of Escherichia coli adenylate kinase. Biochemistry 35, 6100–6106.

190. Schulz, G.E. (1992). Induced-fit movements in adenylate kinases. Faraday Discuss. 93, 85–93.

191. Moore, K.J. & Lohman, T.M. (1994). Kinetic mechanism of adenine nucleotide binding to and hydrolysis by the Escherichia coli Rep monomer. 1. Use of fluorescent nucleotide analogues. Biochemistry 33, 14550–14564.

192. Chao, K. & Lohman, T.M. (1990). DNA and nucleotide-induced conformational changes in the Escherichia coli Rep and helicase II (UvrD) proteins. J. Biol. Chem. 265, 1067–1076.

193. Hakansson, K., Doherty, A.J., Shuman, S. & Wigley, D.B. (1997). X-ray crystallography reveals a large conformational change during guanyl transfer by mRNA capping enzymes. Cell 89, 545–553.

194. Subramanya, H.S., Doherty, A.J., Ashford, S.R. & Wigley, D.B. (1996). Crystal structure of an ATP-dependent DNA ligase from bacteriophage T7. Cell 85, 607–615.

195. Phillips, R.J., Hickleton, D.C., Boehmer, P.E. & Emmerson, P.T. (1997). The RecB protein of Escherichia coli translocates along singlestranded DNA in the 3’□to 5’□direction: a proposed ratchet mechanism. Mol. Gen. Genet. 254, 319–329.

196. Ali, J.A. & Lohman, T.M. (1997). Kinetic measurement of the step size of DNA unwinding by Escherichia coli UvrD helicase. Science 275, 377–380.

197. Chen, Y.Z., Zhuang, W. & Prohofsky, E.W. (1992). Energy flow considerations and thermal fluctuational opening of DNA base pairs at a replicating fork: unwinding consistent with observed replication rates. J. Biomol. Struct. Dyn. 10, 415–427.

198. Barbara Selisko, Hélène Dutartre, Jean-Claude Guillemot, Claire Debarnot, Delphine Benarroch, Alexander Khromykh, Philippe Desprès, Marie-Pierre Egloff and Bruno Canard Comparative mechanistic studies of de novo RNA synthesis by flavivirus RNA-dependent RNA polymerases Virology, Volume 351, Issue 1, 20 2006, Pages 145-158

199. Claudia M. D'Abramo, Luciano Cellai and Matthias Götte Excision of Incorporated Nucleotide Analogue Chain-terminators can Diminish their Inhibitory Effects on Viral RNA-dependent RNA Polymerases Journal of Molecular Biology, Volume 337, Issue 1, 12 2004, Pages 1-14

200. Niklaus H. Mueller, Changsuek Yon, Vannakambadi K. Ganesh and R. Padmanabhan Characterization of the West Nile virus protease substrate specificity and inhibitors The International Journal of Biochemistry & Cell Biology, Volume 39, Issue 3, 2007, Pages 606-614

201. Leandro R. Jones, Rubén O. Zandomeni and E. Laura Weber A long distance RT-PCR able to amplify the Pestivirus genome Journal of Virological Methods, Volume 134, Issues 1-2, 2006, Pages 197-204

202. Stuart D. Blacksell, Syseng Khounsy and Harvey A. Westbury The effect of sample degradation and RNA stabilization on classical swine fever virus RT–PCR and ELISA methods Journal of Virological Methods, Volume 118, Issue 1, 1 2004, Pages 33-37

203. Herbert Weissenböck, Jolanta Kolodziejek, Karin Fragner, Roland Kuhn, Martin Pfeffer and Norbert Nowotny Usutu virus activity in Austria, 2001–2002 Microbes and Infection, Volume 5, Issue 12, 2003, Pages 1132-1136

204. R. Quadri and F. Negro Are there any subgenomic forms of hepatitis C virus RNA in the liver? Digestive and Liver Disease, Volume 33, Issue 6, 2001, Pages 480-486

205. Peter A. White, Zhengqian Li and William D. Rawlinson Sequence diversity in the 5′-UTR region of GB virus C/hepatitis G virus assessed using sequencing, heteroduplex mobility analysis and single-strand conformation polymorphism Journal of Virological Methods, Volume 83, Issues 1-2, 1999, Pages 91-101

206. Udo Künkel, Marina Höhne, Thomas Berg, Uwe Hopf, Alexader S. Kekulé, Gert Frösner, Georg Pauli and Eckart Schreier Quality control study on the performance of GB virus C/hepatitis G virus PCR Journal of Hepatology, Volume 28, Issue 6, 1998, Pages 978-984

207. Takanobu Kato, Masashi Mizokami, Masakazu Nitta, Makoto Nakamura, Hideki Hiramatsu, Kanji Sugihara, Atsunaga Kato, Motokazu Mukaide and Ryuzo Ueda Acute GB virus C/hepatitis G virus hepatitis preceding aplastic anemia Hepatology Research, Volume 9, Issues 2-3, 1997, Pages 164-171

208. J. E. Whitbya, H. Ni, H. E. Whitby, A. D. Jennings, L. M. Bradley, J. M. Lee, G. Lloyd, J. R. Stephenson and A. D. T. Barrett Rapid detection of viruses of the tick-borne encephalitis virus complex by RT-PCR of viral RNA Journal of Virological Methods, Volume 45, Issue 1, 1993, Pages 103-114

209. Christiane Ramelow, Jochen Süss, Doris Berndt, Michael Roggendorf and Eckart Schreier Detection of tick-borne encephalitis virus RNA in ticks (Ixodes ricinus) by the polymerase chain reaction Journal of Virological Methods, Volume 45, Issue 1, 1993, Pages 115-119

210. Stefanie Seipp, Rafael Wahl, Hubert Mueller, Wolfgang Stremmel, Lorenz Theilmann and Tobias Goeser Sequence analysis of hepatitis GB virus C (GBV-C) isolates from 14 patients Virus Research, Volume 46, Issues 1-2, 2 1996, Pages 81-88

211. Thomas Berg, Ullrike Dirla, Uta Naumann, Hans-Gerd Heuft, Stefan Küther, Hartmut Lobeck, Eckart Schreier and Uwe Hopf Responsiveness to interferon alpha treatment in patients with chronic hepatitis C coinfected with hepatitis G virus Journal of Hepatology, Volume 25, Issue 5, 1996, Pages 763-768

212. Eckart Schreier, Marina Höhne, Udo Künkel, Thomas Berg and Uwe Hopf Hepatitis GBV-C sequences in patients infected with HCV contaminated anti-D immunoglobulin and among i.v. drug users in Germany Journal of Hepatology, Volume 25, Issue 3, 1996, Pages 385-389

213. Mini Kapoor, Luwen Zhang, P. Maruthi Mohan and R. Padmanabhan Synthesis and characterization of an infectious dengue virus type-2 RNA genome (New Guinea C strain) Gene, Volume 162, Issue 2, 11 1995, Pages 175-180

214. Modeller, 2004, URL:http://guitar.rockefeller.edu/modeller/ modeller.html

215. Elisabeth Lampe, Jaqueline M. de Oliveira, João L. Pereira, Felipe L. Saback, Clara F. T. Yoshida and Christian Niel Hepatitis G virus (GBV-C) infection among Brazilian patients with chronic liver disease and blood donors Clinical and Diagnostic Virology, Volume 9, Issue 1, 1998, Pages 1-7

216. A Tagger, F Donato, M L Ribero, R Chiesa, V Tomasoni, G Portera, U Gelatti, A Albertini, M Fasola and G Nardi A case-control study on GB virus C/hepatitis G virus infection and hepatocellular carcinoma. Brescia HCC Study Hepatology, Volume 26, Issue 6, 1997, Pages 1653-1657

217. Gabriele Handschuh and Wolfgang H. Caselmann Bacterial expression and purification of hepatitis C virus capsid proteins of different size Journal of Hepatology, Volume 22, Issue 2, 1995, Pages 143-150

218. M. S. Marin, P. M. De A. Zanotto, T. S. Gritsun and E. A. Gould Phylogeny of TYU, SRE, and CFA Virus: Different Evolutionary Rates in the Genus Flavivirus Virology, Volume 206, Issue 2, 1 1995, Pages 1133-1139

www.ingramcontent.com/pod-product-compliance
Lightning Source LLC
LaVergne TN
LVHW082342200926
848481LV00012B/3419